The Soul of a

HORSE

The Soul of a

HORSE

Life Lessons from the Herd

Joe Camp

Three Rivers Press
New York

Photos in chapters 1, 3, 9, 11, 16, 21, and 27 copyright © Pete and Ivy Ramey

Photos in chapters 2, 4, 6, 7, 8, 10, 12, 13, 15, 18, 20, 22, 23, 24, 26, 28, 29, 30, and 31 copyright © Joe and Kathleen Camp

Photos in chapters 5 and 17 copyright © Laurra Maddock

Photos in chapters 14, 19, and 25 copyright © Ginger Kathrens

Published in the United States by Three Rivers Press, an imprint of the Crown Publishing Group, a division of Random House, Inc., New York.
www.crownpublishing.com

Three Rivers Press and the Tugboat design are registered trademarks of Random House, Inc.

Originally published in hardcover in the United States by Harmony Books, an imprint of the Crown Publishing Group, a division of Random House, Inc., New York, in 2008.

Join-Up® is a registered trademark of Monty and Pat Roberts, Inc.

Library of Congress Cataloging-in-Publication Data
Camp, Joe.
The soul of a horse: life lessons from the herd / by Joe Camp.
1. Horses.   2. Human-animal relationships.   I. Title.
SF285.C24 2008
636.1—dc22   2007043820

ISBN 978-0-307-40686-6

Printed in the United States of America

Design by Chris Welch

10   9

First Paperback Edition

FOR EVERYONE WHO HAS EVER LOVED A HORSE . . .

. . . OR LOVED THE IDEA OF LOVING A HORSE

Often, in the early evening, when the stresses of the day are weighing heavy, I pack it in and head out to the pasture. I'll sit on my favorite rock, or just stand, with my shoulders slumped, head down, and wait. It's never long before I feel the magical tickle of whiskers against my neck, the elixir of warm breath across my ear, a restoring rub against my cheek. I have spoken their language and they have responded. And my problems have vanished. This book is for everyone who has never experienced this miracle.

—Joe Camp

Contents

x   *Contents*

Foreword

J oe Camp will immediately settle your nerves. He will cause you to be comfortable in his presence and very soon you will feel as if you have known him most of his sixty-eight years. He has a warm smile and a shock of white hair covering a head that houses a brain most teachers would classify as extremely fertile.

As I write this, I have known Joe less than a year, but it seems a heck of a lot longer. Of course I knew Benji and have marveled at how those stories were put together. But imagine how inadequate it makes me feel to realize how recently Joe came into horses. The man is a natural when it comes to understanding how animals tick and a genius at telling us their story.

The Soul of a Horse will entertain you while it educates you. It will take you on a journey from the prehistoric horse to the modern-day domesticated partner that we all seek to better understand. Joe looks a bit like a Nevada buckaroo, can converse intelligently with the university professor, and is a Hollywood movie producer all rolled into one amazing human being. *The Soul of a Horse* is a must-read for those who love animals of any species.

—*Monty Roberts, author of the* New York Times *bestseller*
The Man Who Listens to Horses, *award-winning natural trainer*
of championship horses, and creator of the world-renowned
revolutionary equine training technique **Join-Up**

The Soul of a

HORSE

INTRODUCTION

My name is Cash. I am horse.

I have been on this planet for some fifty-five million years. Well, not me personally. My ancestors. It all began in North America, somewhere near what is now called Utah. We hung out and evolved for forty-three million years, then we began to migrate, to South America, and across the Alaskan bridge to Asia, Europe, and Africa. And, eventually, some twelve million years after we left, we were brought back home by the Spanish conquistadors.

We've been through it all. Ice Ages. Volcanic periods. Meteor strikes. Dinosaurs. You name it. And we survived.

We've only been carrying man around for, oh, the last three to four thousand years. We've helped him farm, hunt, travel, and fight his enemies. We were helping man shape world history, winning wars for him, as far back as 1345 BC. We protected kings' dominions in medieval times, carried knights into the Crusades, fought on European battlefields all the way into the early 1900s, and helped conquer and settle the American West.

Throughout these millions of years, many of us have always remained wild and free. Even today, our herds roam free in Australia, New Zealand, Mongolia, France, Africa, the Greek Island of Cephalonia, Abaco in the Bahamas, Sable Island in Nova Scotia, the Canadian West, several states of the American West, Virginia, and North Carolina.

And, until recently, we've done it all pretty much naked and in good relationship with man. But over the past several hundred years things began to change. These changes are actually inexplicable, given that our genetics and history are widely known. You see, we are not cave dwellers. We don't like dark cozy rooms, clothing, iron shoes, heat, or air-conditioning.

Humans seem to like all that. And because they do, they presume we should like it too. But we're movers and shakers. In the wild we'll move ten to twenty miles a day, keeping our hooves flexing and circulating blood, feeding our tiny little stomachs a little at a time, and keeping our own thermoregulatory systems in good working order.

Think about it. Our survival through all those millions of years has built a pretty darned determined genetic system. And an excellent formula for survival. We are what you humans call prey animals, flight animals. We are not predators, like you. We have survived because we freak out at every little thing, race off and don't look back. We are also herd animals. Not just because it's fun to be around our pals, but because there is safety in numbers. And being prey animals, we consider safety just about the most important thing. But our idea of safety is not the same as yours. Our genetic history does *not* understand being all alone in a twelve-by-twelve stall. Even if it's lined in velvet, in a heated barn, it's away from the herd and by no stretch of the emotion or imagination is that a safe haven! Stress is all we get from such an experience.

Stress. Big-time!

Have you ever seen one of us, locked in a stall, pacing . . . pawing . . . swaying . . . gnawing? That horse is saying, *Let me outta here!! I need to move! I need to circulate some blood!*

And about these metal shoes nailed to our feet. Have you ever seen a horse in the wild with metal shoes? I don't think so. There is nothing more important to a prey animal than good feet. And ours have helped us survive for millions and millions of years. Rock-crushing hard and healthy.

But once upon a time, back in medieval days, some king decided he would be safer if he built his castle and fortress up on top of a high hill or mountaintop. He still needed us to fight his wars, and move things and people around, but up there on top of the hill, there were no pastures like down in the valley. So he put us in small holding pens where we had to stand around all day, in our own pee and poop, and guess what happened to our feet. It wasn't the moisture so much as the ammonia. Ate our feet up! So when they'd take us out onto those hard stone roads . . . well, you can imagine.

The king's blacksmith came up with the idea of nailing metal shoes onto our hooves, to keep them from disintegrating when pounding the stony roads. There was a much simpler, healthier solution, but, unfortunately, it escaped the king and his blacksmith. So all the king's men and all the king's horses went down the hill . . . and all the king's peasants, living in the valley, where their horses were out in the field, happy as clams with strong and healthy hooves, saw these shiny, newfangled pieces of metal on the king's horses, and what did they say? *Surely the king knows best! We must have some of those shiny metal things for our own horses!*

And so it went for generations.

You humans are funny that way. And you say *we* follow the herd.

Joe and I have had long discussions about all this and he seems to be getting it. So I can shamelessly recommend what follows. Joe has spent much of his life trying to lure you into the heart and soul of a dog, and now he's trying to lure you into the heart and soul of a horse. For it is there that he first began to comprehend the vast differences between us and you, and the kind of thinking that can bridge that gap and bind us together in relationship. My herd mates and I have taught him well. And, believe it or not, the philosophy behind everything he has learned doesn't apply to just horses but to how you humans approach life as well. So whether or not you have a relationship with a horse, I think you'll find this journey of discovery fascinating.

I did.

And I already knew the story.

1

The Herd

The wind was blowing out of the east, which made the beast uneasy. It wasn't normal. And anything that wasn't normal made him uneasy. A stray sound. A flutter of a branch. The wind coming from the east.

But there was a scent on this wind. A familiar scent. One embedded in the big stallion's being for millions of years. He spun on his heels and sure enough, there it was, easily within sight, apparently not realizing the wind had shifted. The stallion screamed to the matriarch, who wheeled in flight.

Like one, the herd followed, racing away at lightning speed, the great stallion bringing up the rear. They ran without looking back for just over a quarter of a mile before the leader slowed and turned.

The predator, a small female cougar, had tired. She had been betrayed by the east wind. The horses had gotten away early, and now she was turning back.

The stallion's senses had saved them this time. The entire herd was alive and well because those very senses had helped their ancestors survive for some fifty-five million years. Prey, not predator, the horse must suspect everything. Every movement. Every animal. Every smell. Every shadow. All are predators until proven innocent. By taking flight, not staying to fight, they survive.

And by staying together. Always together.

How well the big stallion knew this. He had watched his mother, in her old age, lose this very special sense and drift away from the herd. It was excruciating. His responsibility was the herd. To keep them together, and moving. But his mother's screams in the distance would live with him forever.

The matriarch began to lick and chew, a sign that she was relaxing, that all was well. The stallion took her signal, and one by one, the herd began to graze again, nipping at the random patches of grass and the occasional weed. But they wouldn't stay long. The matriarch would see to it. She would move them almost fifteen miles this day, foraging for food and water, staying ahead of wolves and cougars. And keeping themselves fit and healthy.

2

The Student

I remember that it was an unusually chilly day for late May, because I recall the jacket I was wearing. Not so much the jacket, I suppose, as the collar. The hairs on the back of my neck were standing at full attention, and the collar was scratching at them.

There was no one else around. Just me and this eleven-hundred-pound creature I had only met once before. And today he was passing out no clues as to how he felt about that earlier meeting, or about me. His stare was without emotion. Empty. Scary to one who was taking his very first step into the world of horses.

If he chose to do so this beast could take me out with no effort whatsoever. He was less than fifteen feet away. No halter, no line. We were surrounded by a round pen a mere fifty feet in diameter. No place to hide. Not that he was mean. At least I had been told that he wasn't. But I also had been told that anything is possible with a horse. He's a prey animal, they had said. A freaky flight animal that can flip from quiet and thoughtful to wild and reactive in a single heartbeat. Accidents happen.

I knew very little about this horse, and none of it firsthand. Logic said do not depend upon hearsay. Be sure. There's nothing like firsthand knowledge. But all I knew was what I could see. He was big.

The sales slip stated that he was unregistered. And his name was Cash.

But there was something about him. A kindness in his eyes that betrayed the vacant expression. And sometimes he would cock his head as if he were asking a question. I wanted him to be more than chattel. I wanted a relationship with this horse. I wanted to begin at the beginning, as Monty Roberts had prescribed: Start with a blank sheet of paper, then fill it in.

Together.

I'm not a gambler. Certainty is my mantra. Knowledge over luck. But on this day I was gambling.

I had never done this before.

I knew dogs.

I did not know horses.

And I was going to ask this one to do something he had probably never been asked to do in his lifetime. To make a choice. Which made me all the more nervous. What if it didn't work?

What if his choice was not *me*?

*　　*　　*

I WAS IN that round pen because a few weeks earlier my wife, Kathleen, had pushed me out of bed one morning and instructed me to get dressed and get in the car.

"Where are we going?" I asked several times.

"You'll see."

Being the paranoid, suspicious type, whenever my birthday gets close, the ears go up and twist in the wind.

The brain shuffled and dealt. Nothing came up.

We drove down the hill and soon Kathleen was whipping in at a sign for the local animal shelter.

Another dog? I wondered. We have four. Four's enough.

She drove right past the next turn for the animal shelter and pulled into a park. There were a few picnic tables scattered about. And a big horse trailer.

The car jerked to a stop and Kathleen looked at me and smiled. "Happy birthday," she said.

"What?" I said. "What??"

"You said we should go for a trail ride sometime." She grinned. "*Sometime* is today."

Two weeks later we owned three horses.

We should've named them Impulsive, Compulsive, and Obsessive.

OUR HOUSE IS way out in the country and it came with a couple of horse stalls, both painted a crisp white, one of them covered with a rusty red roof. They were cute. Often, over the three years we had lived there, we could be found in the late afternoon sitting on our front porch, looking out over the stalls, watching the sun sink beneath the ridge of mountains to the west. One of us would say, "Those stalls surely seem empty." Or "Wouldn't it be nice if there were a couple of horses ambling back and forth down in the stalls?"

Like a postcard.

A lovely picture at sunset.

With cute horse stalls.

Lesson #1: Cute horse stalls are not adequate reason to purchase three horses.

Never mind the six we own now.

We had no idea what we were getting into. Thank God for a chance meeting with Monty Roberts. Well, not a *real* meeting. We were making the obligatory trip that new horse owners must make to Boot Barn when Kathleen picked up a *California Horse Trader*. As we sat around a table watching the kids chomp cheeseburgers, she read an article about Monty and passed it over to me. That's how I came to find myself in a round pen that day staring off our big new Arabian.

Monty is an amazing man, with an incredible story. His book *The Man Who Listens to Horses* has sold something like five and a half million copies and was on the *New York Times* bestseller list for fifty-eight weeks! You might know him as the man who inspired Robert Redford's film *The Horse Whisperer*. I ordered his book and a DVD of one of his Join-Up demonstrations the minute I got home, and was completely blown away. In the video, he took a horse that had never had as much as a halter on him, never mind a saddle or rider, and in thirty minutes caused that horse to *choose* to be with him, to accept a saddle, and a rider, all with no violence, pain, or even stress to the horse!

Thirty minutes!

It takes "traditional" horse trainers weeks to get to that point, the trainers who still tie a horse's legs together and crash him to the ground, then spend days upon days scaring the devil out of him, proving to the horse that humans are, in fact, the predators he's always thought we were. They usually get there, these traditional trainers, but it's by force, and submission, and fear. Not trust or respect.

Or choice.

In retrospect, for me, the overwhelming key to what I saw Monty do in thirty minutes is the fact that the horse made the decision, the choice. The horse chose Monty as a herd member and leader. And from that point on, everything was built on trust, not force. And what a difference that makes.

And it was simple.

Not rocket science.

I watched the DVD twice and was off to the round pen.

It changed my life forever.

This man is responsible for us beginning our relationship with horses as it *should* begin, and propelling us onto a journey of discovery into a truly enigmatic world. A world that has reminded me that you cannot, in fact, tell a book by its cover; that no "expert" should ever be beyond question just because somebody somewhere has given him or her such a label. That everybody and everything is up for study. That logic and good sense still provide the most reasonable answers, and still, given exposure, will prevail.

My first encounter with this lesson was way back when I was making the original Benji movie, my very first motion picture.

On a trip from Dallas to Hollywood to interview film labs and make a decision about which one to use, I discovered that intelligent, conscientious, hardworking people can sometimes make really big mistakes because they don't ask enough questions, or they take something for granted, or, in some cases, they just want to take the easiest way. In this case it was about how our film was to be finished in the lab, and my research had told me that a particular method (we'll call it Method B) was the best way to go. Everyone at every lab I visited, without exception, said, Oh no, no. Method A is the best way. When asked why, to a person, they all said, Because that's the way it's always been done! In other words, Don't rock the boat.

Not good enough, said I. My research shows that Method B will produce a better finished product, and that's what we want.

Finally, the manager of one of the smaller labs I visited scratched his head and said, "Well, I guess that's why David Lean uses Method B."

I almost fell out of my chair. For you youngsters, David Lean was the director of such epic motion pictures as *Dr. Zhivago* and *Lawrence of Arabia*. I had my answer. And, finally, I knew I wasn't crazy.

It's still a mystery to me how people can ignore what seems so obvious, so logical, simply because it would mean *change*. Even though the change is for the better. I say look forward to the opportunity to learn something new. Relish and devour knowledge with gusto. Always be reaching for the best possible way to do things. It keeps you alive, and healthy, and happy. And makes for a better world.

Just because something has *always been done* a certain way does not necessarily mean it's the best way, or the correct way, or the healthiest way for your horse, or your relationship with your horse, or your life. Especially if, after asking a few questions, the traditional way defies logic and good sense, and falls short on compassion and respect.

The truth is, too many horse owners are shortening their horses' lives, degrading their health, and limiting their happiness by the way they keep and care for them. But it doesn't have to be that way. Information is king. Gather it from every source, make comparisons, and evaluate results. And don't take just one opinion as gospel. Not mine or anyone else's. Soon you'll not only feel better about what you're doing, you'll do it better. And the journey will be fascinating.

We were just a year and a half into this voyage with horses when these words found their way into the computer, but it was an obsessive, compulsive year and a half, and the wonderful thing about being a newcomer is that you start with a clean plate. No baggage. No preconceptions. No musts. Just a desire to learn what's best for our horses, and for our relationship with them. And a determination to use logic and knowledge wherever found, even if it means exposing a few myths about what does, in fact, produce the best results. In short, I'll go with Method B every time.

CASH WAS PAWING the ground now, wondering, I suspect, why I was just standing there in the round pen doing nothing. The truth is I was reluctant to start the process. Nervous. Rejection is not one of my favorite concepts. Once I started, I would soon be asking him to make his choice. What if he said no? Is that it? Is it over? Does he go back to his previous owner?

I have often felt vulnerable during my sixty-eight years, but rarely *this* vulnerable. I really *wanted* this horse to choose me.

What if I screw it up? Maybe I won't do it right. It's my first time. What if he runs over me? Actually, that was the lowest on my list of concerns because Monty's Join-Up process is built on the language of the horse, and the fact that the raw horse inherently perceives humans as predators. Their response is flight, not fight. It's as automatic as breathing.

Bite the bullet, Joe, I kept telling myself. Give him the choice.

I had vowed that this would be our path. We would begin our relationship with every horse in this manner. Our way to true horsemanship, which, as I would come to understand, was not about how well you ride, or how many trophies you win, or how fast your horse runs, or how high he or she jumps.

I squared my shoulders, stood tall, looked this almost sixteen hands of horse straight in the eye, appearing as much like a predator as I could muster, and tossed one end of a soft long-line into the air behind him, and off he went at full gallop around the round pen. Just like Monty said he would.

Flight.

I kept my eyes on his eyes, just as a predator would. Cash would run for roughly a quarter of a mile, just as horses do in the wild, before he would offer his first signal. Did he actually think I was a predator, or did he know he was being tested? I believe it's somewhere in between, a sort of leveling of the playing field. A starting from scratch with something he knows ever so well. Predators and flight. A simulation, if you will. Certainly he was into it. His eyes were wide; his nostrils flared. At the very least he wasn't sure about me, and those fifty-five million years of genetics were telling him to flee.

It was those same genetics that caused him to offer the first signal. His inside ear turned and locked on me, again as Monty had predicted. Cash had run the quarter of a mile that usually preserves him from most predators, but I was still there, and not really seeming very predatory. So now, instead of pure reactive flight, he was getting curious. Beginning to *think* about it. Maybe he was even a bit confused. Horses have two nearly separate brains. Some say one is the reactive brain and the other is the thinking brain. Whether or not that's true physiologically, emotionally it's a good analogy. When they're operating from the reactive side, the rule of thumb is to stand clear until you can get them thinking. Cash was now shifting. He was beginning to think. *Hmm, maybe this human is not a predator after all. I'll just keep an ear out for a bit. See what happens.*

Meanwhile, my eyes were still on his eyes, my shoulders square, and I was still tossing the line behind him.

Before long, he began to lick and chew. Signal number two. *I think maybe it's safe to relax. I think, just maybe, this guy's okay. I mean, if he really wanted to hurt me, he's had plenty of time, right?*

And, of course, he was right. But, still, I kept up the pressure. Kept him running. Waiting for the next signal.

It came quickly. He lowered his head, almost to the ground, and began to narrow the circle. Signal number three. *I'll look submissive, try to get closer, see what happens. I think this guy might be a good leader. We should discuss it.*

He was still loping, but slower now. Definitely wanting to negotiate. That's when I was supposed to take my eyes off him, turn away, and lower my head and shoulders. No longer a predator, but assuming a submissive stance of my own, saying, *Okay, if it's your desire, come on in. I'm not going to hurt you. But the choice is yours.*

The moment of truth. Would he in fact do that? Would he make the decision, totally on his own, to come to me? I took a deep breath, and turned away.

He came to a halt and stood somewhere behind me.

The seconds seemed like hours.

"Don't look back," Monty had warned. "Just stare at the ground."

A tiny spider was crawling across my new Boot Barn boot. The collar of my jacket was tickling the hairs on the back of my neck. And my heart was pounding. Then a puff of warm, moist air brushed my ear. My heart skipped a beat. He was really close. Then I felt his nose on my shoulder . . . the moment of Join-Up. I couldn't believe it. Tears came out of nowhere and streamed down my cheeks. I had spoken to him in his own language, and he had listened . . . and he had chosen to be with me. He had said, *I trust you.*

I turned and rubbed him on the face, then walked off across the pen. Cash followed, right off my shoulder, wherever I went.

Such a rush I haven't often felt.

And what a difference it has made as this newcomer has stumbled his way through the learning process. Cash has never stopped trying, never stopped listening, never stopped giving.

Is this any way to begin a relationship with a horse?

Why would you do it any other way?

3

The Language

The big palomino stallion was anxious to leave, but the matriarch of the herd was scolding a young colt. And the time it took
must be honored, the discipline meted out, or the colt would
grow up a selfish renegade, of no use to the herd, and would most
likely wind up prey to a cougar or a wolf.

After the earlier run, the colt had been feeling his oats, adrenaline and testosterone pumping, and he had snapped and kicked
at a couple of foals half his age. He hadn't really meant any harm,
but it was unacceptable and dangerous behavior in the herd and
had to be dealt with. The mare had squared up on him, back rigid,
ears pinned, and eyes squarely on his. He knew exactly what she

meant, and he now stood alone, well away from the herd. Alone was the scariest place for a herd member to be. Without the protection of the herd, a pack of wolves could easily have their way with him. Before he would be allowed to return, however, he would have to demonstrate his penitence; the mare would eventually swing her back to him and relax, saying the apologies were accepted. He could rejoin the group.

The dominant mare, the matriarch, is the leader of the herd. Usually one of the more mature horses in the group, she serves as disciplinarian, dictates when and where the herd will travel, has the right to drink first from watering holes, and always claims the best grazing. The stallion is the guardian and protector. And the sire of every foal.

The great palomino was taking this quiet opportunity to wander through the herd and check his subjects after the run. It had surely been good exercise, and a sniff here and a look there confirmed for him that there had been no injuries. The steep rocky terrain had conditioned their hooves and legs into appendages of steel. Their daily movement kept the blood flowing and the muscles toned. They were indeed a hearty bunch. But they had to be, for being so was their only defense.

The stallion scanned the horizon, turning a full circle. The sun was low in the west and sometimes caused objects to become mere dark shapes against the light, difficult to distinguish one from another. But one particular shape on a distant ridge stopped him. It hadn't moved, but didn't really look like a rock or a plant. He sniffed the air, but the wind was still coming from the east, and there was no scent other than the sweet smell of Indian paintbrush on the hillside.

The stallion waited. And his patience paid off. The dark shape moved. Turned. His heart began to pump and his nostrils flared. The most feared predator of all! More dangerous because he came astride one of their own, on a horse, capable of running as fast and as far as the herd itself could run. It was a man!

Over the past year, only two herd members had been lost to cougars or wolves, but five had been lost to man. All emblazoned upon the stallion's memory. Long, withering chases ending with herd members being slung to the ground, legs tied, then whipped and dragged around until there was simply no fight left in them, their bodies and their beings stripped of strength and dignity.

The stallion slid up next to the matriarch, adding his burning stare to hers. Saying to the young colt: *Now!* The recalcitrant colt began to lick and chew, and he lowered his head. The two leaders turned their backs, allowing him to return. The matriarch had also seen the figure on the ridge. She uttered a low guttural call to the herd. She must now determine which way to lead them. Certainly not back toward the cougar. Her instincts told her to go south.

She glanced back at the western ridge. The dark shape was gone. There was no time to waste.

4

The Plot

How did we get here? How is it that we have taken this majestic animal, which is fully capable of keeping himself in superb condition and living a long, healthy, happy life, and turned him into a beast of convenience, trained by pain and fear, cooped up in a small stall most of the time, subjected to a host of diseases caused, in most cases, by us.

One would think that the long history of the horse's value to man—as beast of burden, draft animal, riding animal, and companion—would have stirred such a thorough knowledge of his needs that he would have a better, healthier, longer life in our care than he ever could have in the wild. But, in most cases, the exact opposite is true.

According to Dr. Hiltrud Strasser, noted veterinarian, researcher, and author, horses in the care of man have a life expectancy that is, for the most part, only a fraction of that of their wild-living counterparts. Usually because of problems with their locomotor organs. In other words, lameness.

Issues with their feet.

Caused by wearing metal shoes. And standing around all day in a tiny box stall.

Is that a surprise? It was to me. A big one. And it propelled me onto a journey of discovery that quite simply upended everything I thought I knew, and virtually everything I was being told by the experienced and the qualified.

What I discovered was that most humans who own horses have no idea about what's at stake—or what the alternatives are. They're just doing what they've been told to do with no concept that they are causing emotional and physical stresses that depress and break down their horse's immune system, cause illness and disease, and shorten life. And, in so many instances, prevents any kind of real relationship between horse and man.

Dr. Strasser is emphatic that, no matter what you've heard to the contrary, the horse living in the Ice Age, the present-day wild horse, and the high-performance domestic breeds of today are all anatomically, physiologically, and psychologically alike. They all share the same biological requirements for health, long life, and soundness. In other words, we could be not only making the horse's life as good as it is in the wild, but also making it better. At least as healthy. And happier!

Why aren't we?

And what can be done about it?

Finding answers to these questions became the mission. The discoveries were mind-boggling, the solutions remarkably uncomplicated, more often than not involving little more than a willingness to change. A willingness that, bewilderingly, all too often wasn't going to happen.

Leaning on the fence next to me, elbows propped on the top rail, was a true cowboy. Gnarled and weathered, crusty as they come, and a likable sort. Full of tales and experiences. He must've been near my age and had been riding since he was old enough to hold on. I actually paused long enough to absorb the moment, me with my Boot Barn boots and new straw hat, right there in the thick of it. Me and him. Cowboys.

Then he spoke for only the third time since Mariah had come out of her stall, and the reverence I was feeling cracked and shattered like the coyote in a Road Runner cartoon.

Mariah was a cute little Arabian mare that the cowboy had for sale. Kathleen and I were still looking for the right horse for her. The cowboy had watched me earlier in Mariah's stall, just hanging out, waiting for her to tell me it was okay to put on the halter. She never did. The cowboy had asked, "Do you want me to catch her?"

It made me uneasy, but I said, "No thanks." It was that thing about choice again. Trying not to seem so much like a predator by racing into the stall and slapping the halter on first thing, horse willing or not. But I couldn't push away the feeling of embarrassment. Even incompetence. As if I were being challenged. I knew I could corner her and catch her. The stall wasn't that big. But I was attempting to stir some sort of relationship. Not my will over hers, like it or not. Finally, I took her willingness to just stand still as an offer, and I slipped the halter over her head. She made no move to help. I rubbed her forehead. Then her shoulders, belly, hips, and again her face. She twitched, and pulled away, showing no warmth whatsoever.

I led her into the cowboy's arena and turned her loose. It was a small arena, but too large for a real Monty Roberts kind of Join-Up. Still, I had to try. I wanted to see if I could break through the iciness. When I unsnapped the lead, she took off like I was the devil himself, galloping full stride around and around and around. For the most part, I just stood there, doing nothing, mouth agape.

After several minutes, the cowboy asked again, "Do you want me to catch her?"

"No, it's okay," I mumbled, feeling like I was the one on trial, not Mariah.

And she continued to run. I made a couple of token tosses of the lead line, but they were quite unnecessary. She ran on for a good seven or eight minutes with no apparent intention of stopping. I was getting dizzy. Finally, I quit circling with her, turned my back to the biggest part of the arena, dropped my shoulders, and just stared at the ground.

And on she ran. Around and around. I felt the cowboy's eyes on me, probably saying: *What kind of an idiot are you? Get a grip and catch the horse!*

I was running out of will. But Mariah wasn't running out of gas. I was ready to give up when quite suddenly she jolted to a halt. Just like that. Maybe ten or fifteen feet behind where I was standing. I just stood there, staring at the ground. After a moment or two, she took a few steps toward me, then a few more. Monty's advice notwithstanding, I was peeking.

She never did touch me, but she did get within a couple of feet and just stood there. Finally I turned to her, rubbed her forehead, and snapped on the lead rope. I wanted to feel pleased, but didn't. It was willingness without emotion. Her eyes were empty. Like those of an old prostitute. *I know the gig. Let's get on with it.*

The cowboy then climbed aboard to demonstrate Mariah's skills. I suspect he was on his best behavior. He didn't appear to be particularly hard on her, but I noticed that his spurs seemed about two feet long and he did use them. She performed cleanly.

Then it was my turn in the saddle. Mariah pretty much did whatever I asked, but, all the while, her lips were pouty and her ears were at half mast. Neither fish nor fowl. Not really showing any attitude, good or bad. Simply not into it. Not caring, one way or another.

Kathleen was next, woman to woman.

That's when I walked back through the gate and propped myself on the fence rail next to the cowboy. And that's when he said, "I've seen some of that natural horse pucky on RFD-TV and I've gotta tell you, the way I look at it, that horse out there is here for one reason. My pleasure. And I'm gonna make sure she damn well understands that."

I think she did. And, now, so did I.

Clinician Ray Hunt opens every clinic or symposium the same way. "I'm here for the horse," he says. "To help him get a better deal." He and his mentor, Tom Dorrance, were the first to promote looking at a relationship with the horse from the horse's viewpoint. Mariah's owner wasn't willing to do that. His question would likely be, What's in it for me? Rather than, What's in it for the horse?

Perspective is everything, I was discovering. And I wanted desperately to change the perspective of the old cowboy. But what did I know? I was a newbie. A novice. Why would the cowboy or anyone else listen? I felt so helpless.

It would get worse.

As Kathleen dismounted, I looked deeply into this horse's eyes. I rubbed her, and the closer I got, the more she would turn her head or step away. I tried to get her to sniff my hand or my nose. That's what horses do when they greet each other. Sniff noses. All six of ours now go straight for the nose when we approach. Blow a little, sniff a little. And we return the greeting. Much nicer than the way dogs greet each other.

I reached out one last time to rub Mariah on the face, and she pulled away. Just enough. I turned to leave and quite without warning she stretched out and nuzzled my hand. Well, maybe it was more of a bump than a nuzzle. But as I turned back to look at her, it became very clear to me that this cute little mare had received everything I had given, she just had no clue what to do with it. Trust had never been part of her experience with humans.

On the ride home, I asked Kathleen, "So . . . what did you think?"

"No," she said flatly.

The silence telegraphed my surprise.

It seems that during Kathleen's ride Mariah had spooked a couple of times at the dogs barking on the far side of the arena. That, plus the lack of any kind of warmth, had done it for Kathleen. Her blink, her first impression, was *no*.

Two weeks before she had been right on the money. I was all wrapped up in a palomino because he was gorgeous, but I was overlooking at least forty-six shortcomings.

"What don't you like?" I had queried.

"Why would you even ask?" she said. And she was right. It was the wrong horse for us.

Kathleen and I had a deal. We would buy no horse that we didn't agree on.

But Mariah was different. I had finally seen a tiny light in the window. Until later I would have no idea how much she had been saying with that one little bump of my hand. How much of a call it was to take her away. Away from the cowboy.

I told Kathleen about the smidgen of connection, trying to open her mind, but it was locked tight. I felt depressed. I was certain this little mare, given the choice of Join-Up, along with time and good treatment, would come around. She would begin to understand what trust was all about. But I dropped the subject and it was very quiet on the long road home.

The next morning as we sat with our cappuccino looking out over the horse stalls, I brought up the subject again. The next morning as well. And the next. I was haunted by that vacant look in Mariah's eyes and the little bump of my hand. A cry for help. Which I believe to this day it was, but probably not as passionate a plea as I was portraying to Kathleen.

Finally, I'm sure just to shut me up, Kathleen said, "If you really feel that strongly about her, go ahead and get her."

She arrived the next day.

I was excited and anxious to get started, confident that the sincerity of my desire and my extensive working knowledge of the Join-Up concept—which I had been practicing almost a full month now—would win over this cute little mare immediately. I took her straight to the round pen.

No deal.

It didn't work.

She ran around and around, just as she had done the day we met. But no signals of any kind were forthcoming. After several minutes, she clearly wanted to stop, but she had not given me an ear. No licking and chewing. Nothing. So I kept her moving, wondering what I might be doing wrong. Perhaps she didn't know the language of the herd. Maybe she had never known a herd.

Doesn't matter, I objected. She's a horse, with fifty-five million years of genetics. It's in there somewhere. Has to be. I was beginning to reel with dizziness as Mariah continued to run circles around me. Finally, I gave up, put her in a stall, and retreated to the house to watch Monty's Join-Up DVD again.

I watched it twice.

If I was making a mistake I couldn't find it.

Maybe Kathleen had been right. Maybe we shouldn't have purchased Mariah.

Maybe she'd had so much bad treatment that she simply couldn't respond to anything else.

Think persistence, I kept telling myself, remembering the story of an Aborigine tribe in Australia who boasted of a perfect record when it came to rainmaking. They never failed to make rain. When asked how they managed to accomplish such a feat, the king simply smiled and said, "We just don't quit until it rains."

Back to the round pen, and more circles.

Two days of circles! Still no "rain." I was determined that she was going to figure this out. But I was also becoming more and more convinced that she might very well have never been exposed to a herd; perhaps she was one of those horses who had spent her entire life in a stall, with no need for her native language. No opportunity to communicate with horses, and no desire to communicate with people like our friend, the cowboy.

Finally, on the third day, there was a breakthrough. Something clicked. After she made eight or nine trips around the pen, her inside ear turned and locked on me. Then came the licking and chewing. Soon her head dropped and she began to ease closer. I let her stop, turned my back, and lowered my shoulders. Nothing happened for several minutes and I was about to send her off again when suddenly she walked up to me and stood, nose to shoulder. Not sniffing, like Cash had done. But at least she had touched me. Of her own choice. And now she was just standing, instinct in control, but with no apparent understanding as to why.

It was enough. I was grinning from ear to ear.

I turned and rubbed her forehead, and this time she didn't pull away. As I walked across the pen, she followed, right off my shoulder, making every turn I made. I gave her a good rubbing all over. Belly, back, hindquarters, everywhere. And I blew in her nose and sniffed. She didn't respond, but she didn't move away either. I could almost see the wheels turning. *Do I know this greeting? Why's he doing that? I don't hate it really, but I'm not sure what it means. It does seem familiar.*

Somewhere, deep down in her brain, her genetics were finally bubbling to the surface, freed at last from the perspective of the old cowboy.

The next morning when I went down to the stables to feed and muck, I realized for the first time how completely the Join-Up process had transformed Mariah. She was a different horse, waiting by her stall gate, head stretched toward me, and she didn't move until I came over and gave her a sniff and a rub. A scratch under her jaw at the bend of the neck was her favorite. It became ritual. Every morning. And I dared not ignore her or she would scold me with a soft whinny or a snort. And then pull away when I finally came over, just for a moment, to let me know I had been naughty.

The simple act of giving her the choice of whether or not to be with me, of viewing all of her issues from her perspective, not from mine, had changed everything.

The new Mariah is as affectionate as Cash, as willing and giving, as anxious to see us . . . and until Skeeter came along, she was Kathleen's favorite.

I can't help but wonder what the old cowboy would think if he knew that Mariah had learned what it means to trust.

5

Raison d'Être

The herd was tiring. The big mare could sense it. They had run for quite some time. But it was working. She had discovered the concept quite by accident.

The last time man had come after her herd, there were only two choices: a box canyon where they had gone once before and been trapped or the wide-open plains. She had chosen the wide-open run and the pursuers had finally given up. To the matriarch's surprise. Was her herd that much better than the horses chasing them? She had no way of knowing how living with man might affect them.

Whatever it was, it was also working this time. The herd's pursuers were slowing down and dropping off.

As they ran on, the stallion was becoming concerned. Would the very youngest among his herd be able to keep up the pace? But just as one little filly began to drop behind, the last of the pursuing horses stumbled, almost losing his rider, and they turned back, quitting the chase.

The mare slowed to an easy canter, but kept moving for a while before bringing the herd to a halt. She scanned the horizon in all directions. Night was falling. Would the pursuers be back tomorrow? Would they try to sneak up on them during the night? She would be prepared.

She showed the herd that she was relaxed, at least for the moment, and they began to graze. All except a small sorrel mare who was designated sentry. She would be watching, always watching. And listening, aware of everything.

The big palomino shook off the chill of night air settling upon his sweaty body and wandered through the herd to check for injuries. As before, there were none. Not even among the young. He was proud of that. His herd was well conditioned, and their feet were rock solid.

A harsh nicker drew his attention. The young colt was at it again and the matriarch was dealing with him. The stallion would leave it to her. He was tired, and perhaps a bit of sleep would be good.

Good sleep. Not the standing kind.

He made his way into the middle of the herd and eased down onto the ground. The herd would keep him surrounded until he woke.

6

The Starting Gate

"That horse is mean. He was born mean!"

It was a trainer at the local horse club speaking.

I didn't believe him.

There was a time when I would have. But Monty and well-known natural horsemen like Pat Parelli, John Lyons, renowned equine vet Dr. Robert M. Miller, and a host of others say no horse is born mean. They are *made* mean by humans, usually because the human doesn't understand or doesn't want to deal with the concept that the horse is a flight animal. Flight is so embedded in their genetic makeup that reaction is automatic. Any sound, or smell, or flicker of movement that is unfamiliar can cause them to erupt. *React first, ask questions later.*

And some folks read this as being bad tempered. Mean.

Mariah once freaked out over a squirrel in the brush and leaped three feet sideways, straight into Kathleen, knocking her to the ground. She wasn't being mean. She wasn't trying to hurt Kathleen. She was probably trying to jump into her pocket. *Save me, Mommy!* Because Kathleen had proven herself a good and trusted leader.

The trainer mentioned above would've likely beaten the horse, without even considering the fact that she was simply afraid, and reacting in the way that horses have reacted, automatically, for millions of years. A beating would cause more reaction and one thing would lead to another. Soon the horse would come to believe that humans are mean. Predators. And anytime one gets near, it will most likely mean pain. So they become even more fearful. They try to take flight, and if they cannot, they resort to a last-ditch attempt to protect themselves from what they are certain is about to come.

Kathleen kept her adrenaline down. Which, in turn, calmed the horse. She picked herself up and rubbed the little mare, letting her know that she was safe, that nothing was going to harm her. And she made a mental note to always do what renowned Australian clinician Clinton Anderson preaches: Don't stand right next to a horse until she's well along in her training and desensitization! In other words, stay out of harm's way and be prepared. Then slowly teach the horse. Desensitize the horse to whatever makes her afraid. Let her know that in your presence she's safe. All the while, teaching the horse to respect your personal space, and teaching her to focus; to get back to thinking instead of reacting, keeping ever in mind that, like us, horses have different makeups. Some are *very* sensitive, and some are not. Some are more freaky than others. Some learn fast, some slowly. Some are more mischievous than others. But if they have made the choice, on their own, to trust and be with you, with that comes a willingness to learn and to follow your lead.

They need only to understand what it is they should be learning. Which puts the ball squarely in our court. How do we become clear communicators? Without domination, intimidation, meanness, cruelty, or pain.

Monty Roberts has scores of tales about horses no one could go near, horses most folks would place well inside the *mean* category. But they are now happy, well-adjusted partners. One in England took three full days to come around, but come around he did. This was a horse who had obviously been badly abused somewhere in his history and had decided that all humans were agents of pain. Monty convinced him otherwise.

Yet with such positive results coming from so many different directions, why are we still where we are today, with so many owners of horses living in the dark ages? The reason, I believe, is that most people do not begin at the beginning. They want to start halfway around the track, instead of in the starting gate.

I now have a horse. I want to do something with it. Go riding. Compete. Something!

We humans are in such a hurry that there's no time to build a relationship. To learn to communicate. To gain and give understanding. To walk in the horses' boots, so to speak.

To begin at the beginning.

The beginning for us was our discovery of Monty Roberts and his Join-Up process.

And Kathleen's fear.

She was petrified, and I had no idea.

That birthday trail ride was not something she was looking forward to. It was a gift for me. I was suffering from the so-so results of the last Benji movie and she had wanted to find something for my birthday that would be a diversion and make me smile. That's the way she is.

But when her fears began to creep out of the closet, I became even more committed to making sure our new horses were safe and our relationship with them well founded. Begin at the beginning. Take whatever time it takes.

Clinician John Lyons says that there is a real reason for fear: "Fear is recognition of loss of control, and it subsides when control returns."

That's why so many of the DVDs and books begin with the art of gaining willing control of the horse's body parts. Moving them this way and that. Backward, forward, and sideways. Both on the ground and in the saddle. That control buys safety, and respect from the horse because you are speaking the language of the herd. Who moves who.

But looking back on Kathleen's fear, I would adjust John Lyons's phrase to read "Fear subsides when you *believe* that control has returned." Because all too often I would watch Kathleen do a splendid job of controlling her horse, only to find out later that she didn't *believe* she was in control. She was going through the motions, doing what the DVDs and books said, but she didn't believe she was actually leading the horse. There was no connection. She was merely executing a learned exercise without understanding the point of the exercise.

My fear threshold was much higher than Kathleen's, probably because at some unconscious level I was actually relating to the horses. Perhaps embedded from my decades of work with dogs, of living inside their hearts and souls trying to draw you in. When Cash walked up behind me and touched my shoulder in that first Join-Up, there was an emotional exchange. I gave absolute trust to him when I turned my back. He returned it when he touched me on the shoulder. I know now there was no such exchange when Kathleen did her first Join-Up. She went through the motions but was petrified that the horse might walk up behind her and knock her down. There was no offering of trust. She knew it, and, unfortunately, the horse knew it as well. As wacky as it might sound to anyone who hasn't been there, they do read these things. Unerringly. Horses will never be dishonest with you, and they will always know when you are dishonest with them.

But I wouldn't learn most of this until further down the line. At the time I was just beginning to realize that something was amiss, for both of us. Kathleen was having trouble giving trust to receive it, so she was unable to actually *believe* she was in control. I knew I was in control, but had no idea why. And when logic is removed from any equation, it makes me crazy.

Our growing library of books and DVDs all said to begin at the beginning, which meant standing in the arena teaching my horse to back up or move sideways. Or come to me. These exercises would give me control, said the DVDs. And once I had complete control over how, where, and when the horse moves, I would then have a safe horse. And only then should I climb aboard.

But I wanted to know why.

I was also anxious to take the next step with Cash. After Join-Up, he was now looking to me for leadership, so off we went to the arena.

I hear we learn by our mistakes.

One of the training DVDs had spelled out three different ways to teach backup:

See Cash back up, Method One.

See Cash back up, Method Two.

See Cash back up, Method Three.

Why, I wondered, did I need three? Especially here, beginning at the beginning. One method would've been quite enough to confuse both of us this first time out.

See Joe look like a circus clown.

Clumsy and awkward do not adequately describe the moment. I had Cash's lead rope in one hand and a three-foot-long Handy Stick in the other. A Handy Stick is a plastic rod used to extend the length of one's arm so that, hopefully, one can stand back far enough to avoid the kind of knockdown Kathleen got to experience. The stick, sold, of course, by one of the DVD trainers, is not to be used for discipline, only for guidance. According to this particular DVD, I was supposed to be doing one thing with the lead rope and another with the stick.

It was like trying to rub circles on your belly with one hand while patting your head with the other.

I felt like an idiot.

Those droll cocks of the head and quizzical looks from Cash were coming at me like machine-gun fire. I expected him to burst out laughing any minute. I was clearly not getting through. But something else was bothering me, something beyond beginner's clumsiness. Cash and I had bonded just a few days before in the round pen, and this exercise was not strengthening that bond.

I was trying to learn a specific task, or, rather, trying to *teach* a specific task.

Or both, I suppose.

But what I wanted most was to understand what made this huge wonderful beast tick: how he learns, how I could communicate clearly with him, what it meant to him when I did this with the rope or that with the stick. Only then could I better figure out how to get Cash to understand what he clearly wasn't understanding on this particular day. I was trying to follow a DVD's instruction, move by move, when what I felt I should've been doing was experiencing this from his end of the rope.

It wasn't long before Cash sent me straightaway back to the books and DVDs, which, I soon discovered, had no intention of teaching me how to understand Cash until I first knew how to back him up. And move him sideways. And so on.

Truth be told, I actually went back to determine whether the stick was supposed to be in the right or left hand. As if it made a difference to Cash.

But I found myself skipping ahead, to the end of the series. Looking for some conceptual meat. Then to the end of the next series. Racing past the task-based learning. Searching for comprehension, meaning. Something that would connect the dots. What I found is that all these programs pretty much never get there until the end.

How backward, I thought.

Now that you've come this far, we're going to teach you why all these tasks we've taught you work. We're going to show you how to understand the horse so you can figure things out for yourself.

But I didn't want to wait. Not that the tasks don't have good and proper purpose. It's just that they would mean so much more to both of us, to me and the horse, if I understood why he was getting it, and why he wanted to. Learning, then, would surely happen so much faster. His *and* mine.

The early lessons in the books and DVDs never said: Before you start this program, go spend a few days out in the pasture just watching the interaction of the herd. Make note of how the smallest of gestures, when delivered accurately, can get the desired result.

Wow! Why wasn't that up front?

I went immediately to the pasture. And watched.

Again and again.

The DVDs didn't explain, in the early lessons, that when a leader horse swells up and pins her ears and moves toward a follower's butt, it means move that butt. Now! And that such a move doesn't mean *I don't like you.* Or *I want you out of my pasture.* It simply means, *I am the leader here and I want you to move your butt over.* That's it. A few minutes later the same two horses will be huddled next to each other, head to tail, swishing flies out of each other's faces.

This is a difficult concept for humans to grasp. We are such emotional beings. We don't like to hurt another's feelings. Usually. So it's hard for us to realize that with horses, such behavior is simply leadership in action. And is actually building respect for the leader.

It was important for me to learn that the horse was not going to think less of me if I swelled up like a predator, pinned my ears, and pointed at his hindquarters. He would actually think more. *Hey, this guy knows the language. Cool. I respect that.*

And then it struck me: These horses accepted me in Join-Up. I'm *supposed* to be part of the herd. It stands to reason that I need to know how to behave like a herd leader.

The hard part was remembering to swell up. And I had trouble pinning my ears. I suppose that's why we have fingers. And eyebrows. Eyebrows are good.

I was beginning to understand that, in effect, we must find a way to be a horse. We shouldn't even try to relate horse behavior and communication to human equivalents. Or even doggie equivalents. Horses are not humans. And they aren't dogs. If you treat a horse like a puppy, you will never be his leader. I'm not saying you shouldn't give your horse a hug or a rub. But a dog will do virtually anything for a hug. A horse will do virtually nothing for a hug. But he will do virtually anything for his respected leader. And he will continually test that leader to see if he or she is still worthy of the title.

It was in the pasture that I learned all this, and began to understand how to be a horse. I had finally found where I was to begin. I was ecstatic.

None of the DVDs had said any of this early enough to suit me. And very few effectively embraced the concept of how a horse learns until well into the program. Simply understanding what *reward* is to a horse made so much difference in the way I approached the task of training. But like learning to get out into the pasture, I had to skip ahead in those DVDs to find it.

Reward for a horse, I finally discovered, true reward, comes from release of pressure.

And with that reward comes learning. Communication. Understanding.

It's as simple as that.

In the wild, when being chased by a cougar, the horse's reward is when the cougar turns back.

Release of the pressure.

And so it is in the herd. When the matriarch disciplines the foal by sending him away from the herd, and pressures him to stay away, it is the release of that pressure, when the foal submits, that is his reward. As the foal begins to understand what it takes to avoid the pressure, he will submit earlier the next time. And, hopefully, not be a bad boy at all the third time.

When a dominant leader says, *Move your butt over*, the instant the follower responds, the leader drops the pressure. The lesson: *If I move my butt when she applies pressure, she will release the pressure and I will no longer feel uncomfortable.*

The next time, that same horse will move his butt sooner. And before long, a simple look from the leader will do the job. No swelling up. No pointed movements. Maybe just a drop of the ears. Or a flick of the head.

And so it is as we teach. It's not so much what we do, but rather the release of pressure the very instant the horse gives even a hint of the desired response. Then, depending upon the horse, it usually doesn't take long to reach the conclusion: *Oh, I get it. If I move over when Joe does that, he releases the pressure, so that must be what he wants.*

In effect, this is an extension of the doctrine of choice. *Do I want the pressure or the comfort of no pressure? I think I'll move over and thereby choose no pressure.*

Maybe Kathleen and I are weird, but we agree that having a

thorough understanding of how a horse learns, and how the herd teaches one another, how they receive information and understanding, would provide so much more insight into the training process. And be a richer foundation from which to launch.

Concept-based learning.

This all came together for me one day as I was scanning a DVD and stumbled onto a very simple little exercise with an in-depth conceptual analysis of why it worked. The exercise was simply to get the horse to lower his head when asked. No sticks involved. No arenas. No stumbling around trying to rub my belly while patting my head. Just me and Cash. Up close and personal.

The lesson began with understanding that all horses, by nature, resist pressure. And lean into pressure. When you push on an unschooled horse's hindquarters, the hindquarters will come toward you. Push on his shoulder and he'll lean into you. Pull down on an unschooled horse's halter and he will resist and pull up. That's because the pressure, to him, is actually at the top of the halter. He feels the top strap pushing down. So he pushes into that pressure by lifting his head. These are genetic traits, embedded for survival. When a wolf sinks his teeth into a horse's underbelly, the horse's only chance for survival is to push down, to apply pressure to the wolf. If the horse pulls away from the wolf, he helps the predator rip his belly open.

So how do you get the horse to understand that you want his head to go down? How do you communicate that when he wants to push against the pressure by raising his head?

How would I do it with a dog? How would I get Benji to understand a desired action?

I would reward him with a treat.

And what is reward to a horse? I asked myself with a knowing smile.

Release of pressure. Comfort, I said smugly.

And off I went to gather Monsieur le Cash.

This time it went swimmingly. I applied the slightest of downward pressure to the lead rope. Not trying to pull his head down. Just enough to counter his upward resistance. And I held it. The discomfort to Cash was minimal. Just the pressure of the rope halter. Before long, Cash lowered his head, just enough to release the pressure, and I immediately dropped the lead rope, rubbed him on the forehead, and praised him.

Then we did it again. This time he dropped his head sooner, and went farther down, and I released the rope, as Clinton Anderson says, like it was a hot potato.

Before long, Cash's response was almost instantaneous, dropping his head as much as ten to twelve inches. I pulled out a folding chair and sat down to see if he would drop all the way to my lap. Three sets of pressure and release, and he was there. I could've bridled him from the chair. Granted, Cash is very intuitive. He gets things quickly and is very willing. Others of our horses would take longer. But now they all have learned this task.

The next step was to ask Cash to leave his head down, rather than immediately lift it up upon the release of pressure. To communicate that, when I released the rope, I released it just a little. When he lifted up, he bumped back into pressure from the rope and immediately dropped again.

He was soon staying down until I completely released the rope and said, "Okay. Good boy."

I was grinning from ear to ear.

Not so much because he had done the task, but because I had watched his wheels turn. I had seen the intake of understanding that I was asking for something that was completely counter to his genetics, but because I was a trusted leader, he could respond safely, without worry. Willingly.

And he did.

We tromped up the driveway to the front door of our house and I called for Kathleen.

"Come out! I've gotta show you something!" You would've thought I had found the cure for cancer.

The door swung open and she almost swallowed the plum she was eating. She had never seen a horse at the front door before. Cash was virtually inside, his curiosity working overtime.

I demonstrated Cash's new feat and rattled on about the learning process. The discovery of pressure and release.

"Have you tried that with his ears?" she asked.

Cash had come to us with one rule: *Do not ever touch my ears!*

We had often wondered what might've happened in the past to cause this reaction. I've heard of trainers who have been known to twist an ear to make a horse do or accept something. Whatever it was, we couldn't get close. Couldn't even scratch Cash between the ears.

"Good idea!" I said.

The *pressure*, in this instance, would come from his own fear of humans touching his ears.

I reached slowly up the side of his head toward his ear. He immediately pulled away when I got too close. My hand went with him, staying in position, creating even more pressure, until he stopped and held still for a couple of seconds. Until he was able to realize that it wasn't going to hurt him. Until he relaxed. Then I removed my hand.

It happened. He finally began to think, *This is no big deal.* That bought him a release of pressure. More comfort.

I reached again, a bit farther.

When he didn't retreat, I dropped my hand. The release of pressure sent a message, just as it had when I'd released the halter while teaching him to lower his head.

One more time. Gaining an inch over the time before. And I retreated immediately when he didn't pull away.

And rubbed him on the forehead. *Good boy.*

And so it went, gaining an inch or so with each try. If he pulled away, I'd go with him until he relaxed. It didn't happen often. He was getting the picture.

It took about ten minutes before my fingers were wrapped around the base of his ear, rubbing very gently. Then withdrawing.

I quit for the day, feeling there had been a major breakthrough.

The next day, after maybe twenty minutes of microprogressions, I wound up with my hand wrapped all the way around his ear and my thumb rubbing gently *inside.*

Just amazing.

Approaching the other ear was not quite like starting at square one, but close. By the end of the week, I could rub both ears, inside and out, and today Cash virtually purrs when we do this, leaning into it, saying, *More, more.*

It's truly exciting what a bit of understanding can do.

And patience.

That's the huge lesson Cash, and all the other horses, are teaching me. I've never been accused of having a lot of patience.

Not even a little.

Cash showed me the way.

Again.

Don't start halfway around the track, Joe. Start at the starting gate. Because when faced with an unruly horse who hasn't begun at the beginning, a beast six or seven times your own weight, it's a knee-jerk reaction to attempt to overcome the size relationship with force and dominance.

I remember one of the first times I went on a trail ride. A mere kid, primed with years of cowboy movies, I wanted to let this huge creature know in no uncertain terms who was boss. Understand that this poor horse had probably been doing the same thing, dealing with idiots like me, day in and day out, for longer than I had been alive. But there I was, reins pulled tight, jerking this way and that, kicking his sides, establishing my dominance. Without a single clue.

The real embarrassment is that, decades later, when Kathleen arranged the birthday trail ride, I was doing exactly the same things. Establishing my bossmanship. Looking like I knew what I was doing. Soaking up compliments from the trail leader.

And in a way, I suppose, all of that's fine for the occasional trail rider. Most trail horses know so much more than those who ride them, it's difficult to do too much wrong. They won't let you. They are turning, going, and stopping before you think about it so you don't have to jerk on their mouth or kick them in the side. The years have taught them.

But for the horse owner, there's only one place to begin.

At the beginning.

Stand in the horse's hooves. Study his history. Understand why he is the way he is, and why he acts the way he does.

He's a prey animal.
You mean like a rabbit??
Pretty much, yeah.
But he weighs eleven hundred pounds!!
Yep.

Discover what makes him feel safe. What keeps him healthy. What he wants in a leader. And why.

The following story is not mine. I asked and received permission from Monty Roberts to summarize it here because I feel it's so important to understand what can actually be accomplished. Monty has written an entire book, entitled *Shy Boy*, on the subject. I encourage you to read it.

Monty was asked by the BBC if he thought he could accomplish his Join-Up procedure totally in the wild. Without round pens, without lead lines. Just him and a wild horse. A mustang. He said yes, and a few months later he did just that. With cameras rolling, he joined up with a mustang in the wild, saddled, bridled, and placed a rider on the horse he later named Shy Boy. It took something like thirty-six hours to accomplish this feat. Monty was in the saddle of his own horse for most of that time. An amazing accomplishment. But the most important part of the story is this: A year later the BBC called again and asked Monty what he thought Shy Boy might do if he were returned to his herd. Would he choose to stay with the herd or would he stay with Monty?

Frankly, Monty wasn't sure he wanted to find out. He now loved this horse. But persistence from TV producers convinced him. And, again with cameras rolling, they found the herd and released Shy Boy.

The mustang took one look at the herd and loped off to join them. They were last seen that evening literally racing off into the sunset. Monty stayed awake most of the night. He had lookouts positioned all over the place with radios, watching for the herd. Around nine o'clock the next morning, a radio crackled and blared that the herd was in sight, headed more or less their way. Shy Boy was out front.

At the bottom of the ridge that separated the horses from Monty's encampment, the herd stopped and Shy Boy climbed to the top of the ridge. He stood for quite some time looking first at the herd, then at the camp. Finally, he turned and galloped down the ridge toward the camp, weaving in and out of tall brush, slowing to a trot, then a walk, stopping only when he was nose-to-nose with Monty. I cried like a baby when I read that story. Imagine how you would feel if that was your horse, turned loose to make his own choice, to run free with his herd or come back to you. You would surely know that you had been doing something right.

We were finally able to meet Monty about a year ago when we got together at his ranch to discuss his possible involvement in an upcoming Benji movie. But it was difficult for me to stay focused. This man is an icon in the horse world, and I had read his every book and seen all of his DVDs. I was mesmerized.

Just listening firsthand to Monty speak of his experiences, and twisting and pulling ideas with him, made it a very special encounter. But the highlight of the day—sorry, Monty—was meeting and being able to Join-Up with Shy Boy himself.

Seeing our twins, Allegra and Dylan, then twelve, Join-Up with this famous horse as if they had been doing it all their lives confirmed forever the simplicity and value of the process. They did a much better job than I did. With the master himself barking directions and correcting my body positions, I felt like a bumbling buffoon.

But Shy Boy made up for all of our shortcomings and was having a terrific time showing us the ropes. There was no question that this was one happy horse.

I am quite certain that when he was turned back out into the wild, Shy Boy would never have returned to the cowboy who sold us Mariah. Or the trainer mentioned at the beginning of this chapter.

But he returned to Monty Roberts.

I tell everyone Shy Boy's story. I tell it over and over again.

It's the way it should be. And it doesn't have to be any other way.

It's what happens when you begin at the beginning.

7

To Sleep
Perchance to Dream

By human standards it had been a very short nap, but it was deep REM sleep, and it was all the stallion needed. He had dreamed about the humans on horses who had been chasing the herd. One had gotten close, and the stallion had turned and confronted the horse, who then reared, threw his rider, and raced off with the stallion to rejoin the herd. As it should be.

The herd had encircled him while he was sleeping, protecting their protector. The sentinel was on duty. It's the same anytime any horse lies down. The herd gathers and guards. The sentinel watches and listens. Horses need REM sleep and cannot get it standing up. But on the ground they are more vulnerable to predators, so most horses will not lie down unless guarded by the herd. One of the many reasons why nature never intended horses to live in isolation.

It was dark now, and the golden stallion was up and about, fully refreshed. It was time to change locations. Move away from where they were last seen by man. He nudged the mare and she wandered through the herd, nuzzling and growling. Calling the troops to order. *Time to move on.* And they did.

The herd would move seven miles to the south before stopping again. A small stand of spruce along the edge of a ridge would provide all the cover they needed for others in the herd to get some sleep. The men, if they did come looking, would not find them, would not even be able to see them until morning, and by then they'd be gone, once more on the move.

8

The Wild Horse Model

Our new natural pasture was ready. Not a pasture in the *green grass* sense of the word. This one was dirt and rocks and virtually straight up and down. Very steep. But in those ways it actually matched our research on the *wild horse model*, which, in essence, is an attempt to imitate the way horses live in the wild. And it was the best we could do with the virtually unusable acre and a half behind our house. It was surrounded by an inexpensive electric fence inside a perimeter fence of chain link that was already in place.

And now it was time to try it.

We were worried.

Was it too steep? Would the horses like it? Were there too many rocks and boulders? Would they hurt themselves? Would they all get along in the same pasture?

Scribbles was first.

He's the quiet one. A gorgeous paint, but not long on charisma. He's the one most likely to be found standing in a corner, seemingly motionless, for hours. *Lazy* would be a merciful understatement. He has the best *whoa* of any of our six, because it's his favorite speed. No reins needed. Just sit back a little, then hold on for the screech of tires. *Can we stop now?* is his favorite question. He leads like it's an imposition to ask him to move. *Oh, all right, if you insist, but you have no idea how much effort this is.*

Which is why his first venture into the natural pasture left me with my mouth hanging open in astonishment. As the halter fell away, he spun and was gone like a bullet. Racing, kicking the air, tossing his head, having the best time I'd ever seen him have. This was not a horse I had met before. He went on for a good ten minutes, with me just standing there, grinning like an idiot.

I could imagine that somewhere inside those two brains he was screeching *Whoopee! I'm free! I'm free!* Finally, he trotted back over and in his own begrudging little way said thank-you. That was the beginning of a new way of life for Scribbles and his five herd mates.

Horses in the wild, on average, are healthier and more sound and, under decent conditions, live longer than horses in domestic environments, say Dr. Strasser and natural hoof specialist Jaime Jackson, among others. That doesn't mean we should turn all our horses loose. It means that we should exert every effort to care for them at least as well as they care for themselves in the wild. To pattern their care after the wild horse model rather than after the human or dog model. In effect, to replicate as much of their lifestyle in the wild as is humanly possible.

As the logic of that research sank in, Kathleen and I often found ourselves looking at each other through astonished eyes. Either our discoveries were truly amazing or we were certifiably nuts. How could so many people be so wrong for so long? It simply didn't make any sense.

"But we began the same way," Kathleen said one day. "We were right there, buying into the same things."

She was right. When we began this journey, I'm not even sure we realized that wild horses still existed. We certainly didn't know that they had been around for fifty-five million years. So, like most people, we had given no thought to how they had survived all that time with no assistance from humans. We had never read Dr. Strasser, so we didn't know about her research, which concludes that horses in the wild today can live up to twice as long as the average domestic horse. So, of course, it follows that we had no idea why. We simply had no knowledge of any of it, so how could it apply to us? Which, unfortunately, is the case with most of the folks we've run into.

Now, when we start spewing all this information at some unsuspecting horse owner, eyes widen and jaws drop. And some of them rush off to get away from the weirdos.

But many of them rush off to be pardoned by their horses.

And that makes it worth the effort.

These are the easy ones. The open ones. The ones who see the logic of it all.

The tough ones say, "Oh, that's so wrong. The domestic horse has been bred and cross-bred so many ways that he's not even the same species as the wild horse anymore."

This is unfortunate because the experts tell us that millions of years of genetics could never be wiped out by a few generations of selective breeding. The reason many of these folks truly believe what they're saying is because their horse has had some sort of disorder, like lameness, for so long that they are certain he must be genetically unfit. However, the problem most often lies in his metal shoes, or his tiny stall, or his diet, or stress.

Documented case after case confirms that these unhealthy domestic horses can become healthy again if the source of their ill health is removed, if they're given the opportunity to live as nature intended.

The story of Shy Boy answers the question of whether horses can be happier domestically than in the wild. Without debate. We know from study and firsthand experience that humans can have amazing relationships with horses. They can become part of the herd and be chosen as a leader. But a good herd leader cares for his or her herd in the best possible way. The knowledge we've gobbled up and the experience we've gained overwhelmingly confirms that the best possible way is from the perspective of the horse, not the perspective of the human. Horses want and need to feel safe, to be as healthy as possible, and to live as long and happy a life as they can. And none of that happens when they're away from the herd, motionless in tiny stalls, eating only a couple of times a day.

I was out with the horses one night, thinking I could get the feeding and mucking done before the rain started. The rain that wasn't even supposed to be here in October. But there it was, and wet I was, and cold. The temperature was only in the midfifties, but to me, sopping wet, that was *freezing*.

I looked at our horses, heads down, dripping with water, and I just couldn't stand it. I went for the halters and lead ropes and brought them into their covered stalls. The stalls are open, actually only half-covered, with one solid side facing the usual weather assault, but if we'd had a cozy barn with central heating and warm fuzzy pillows, I'm sure I wouldn't have hesitated to take them right in.

It's difficult for humans, especially when cold and wet, to understand that the horses do *not* feel like we do.

I was traveling with my son, who was looking for investment property in northern Idaho. We had been driven out to a gorgeous ranch with a huge log house, fenced pastures . . . and a spectacular six-stall barn. While my son was kicking tires and asking questions, I was having quite the nice time strolling around this beautiful place, lusting after such a spot, especially the barn. What is it about us humans that makes us want a big barn almost more than we want the horses that would go in it? As I walked down the center aisle, I was struck by how clean it was. Pristine! When the owner happened by, I said, "Do you never use this barn? It's so clean."

"Oh sure," he said. "We use it for hay storage."

"What about the horses?"

"They like to be outside."

"Even in the winter? In the snow?"

"Yep."

We were only twelve miles from the Canadian border. Winters are not warm here. I was amazed.

The owner walked around the barn to show me a lean-to he had built, which was attached to the side of the structure. Just a roof, with divided stalls, to keep the horses separated when eating grain. They had free access to this shed, but never came into it except for their grain. Again, I was amazed. This ran so counter to everything I felt for my horses. We want to think of them as big dogs, and treat them in the same manner.

They aren't big dogs.

Not even close.

Dogs, like humans, are cave dwellers—predators, who feel comfortable taking care of themselves. They run in packs primarily to gain advantage in the hunt, not for safety. Pat Parelli says humans and dogs are most interested in praise, recognition, and material things. Horses are not interested in any of that. Horses are interested in safety, emotional comfort, play, food, and procreation. Praise, recognition, and material things are of no interest to them. At all.

Hard to grasp, isn't it?

But, come on, what harm can it do to show a bit of TLC by storing them away in a nice comfy stall, with central heat and air, a bit of velvet on the walls, and a soft, cushy floor?

A lot of harm. Believe me when I tell you: a lot.

What we humans feel our sweet babies should have is most often exactly the opposite of what they need for health and happiness.

A horse's entire physiology has been built over millions of years to:

- Move a minimum of fifteen to twenty miles a day, on bare hooves
- Be with the herd, physically and thus emotionally safe, unstressed
- Spend sixteen to eighteen hours a day eating . . . from the ground, a variety—continuous uptake in small quantities to suit their small tummies
- Control their own thermoregulatory system, thus controlling their own internal body temperature with no outside assistance, such as heat, blankets, and the like

- Stand and walk on firm, fresh ground, not in the chemical remnants of their own poop and pee . . . nor be breathing the fumes of those remnants, plus the excessive carbon dioxide that accumulates inside a closed structure
- Get a certain amount of unstressed REM sleep, which usually will only happen when surrounded by a herd with a sentry on guard

Dr. Strasser compares the complex interaction of the equine organism in its natural surroundings to a key in a very complex lock. Alter anything on either side—grit in the lock, a corner broken off the key—and the entire system is no longer functional.

Of course, a horse who has been in a controlled temperature environment or who has been wearing a blanket has not been able to grow a winter coat and lacks the stimuli (temperature fluctuations) that trigger and strengthen the activity of the thermoregulatory system. That horse must be introduced to winter gradually. Put him out in the early spring and let his system reacclimate. He'll then be ready for the following winter.

The only area where the horse might need a bit of help is under extremely cold conditions when it's raining. When a horse's coat gets soaked, and it's really cold, and he cannot get out of the wind, his systems might become overloaded. I've found no hard research on this, so I say better safe than sorry. If there is no natural windbreak in the turnout or pasture, provide, perhaps, a covered windbreak where he can stay clear of rain in extremely cold weather, and he'll be fine.

When I was standing out in the cold rain, without a raincoat, feeling sorry for my horses, I didn't want to hear, "Your horses are fine, Joe. Leave them be." It was difficult for me to believe, as miserable as I was feeling, that the horses weren't miserable, too. But the truth is, they weren't. And the things I've been seeing, like the horses on the trip to Idaho, always push me to learn more, to dig, to throw out the marketing-induced guilt of the barn and blanket makers, the "traditional" reasoning, and try to get to the truth. For no other reason than I care for my horses as much as they appear to care for me.

When we take control of one of these lives, when we say, *I will be responsible for this animal, his care and feeding, his health and happiness,* we tacitly promise to give him the very best care that we can. To learn everything we can about the horse, and how to give him the longest and very best life possible. Not the life *we* think he should have because that's what *we'd* like, but the life we *know* is right because we've studied it and *are certain.*

Yet the majority of domesticated horses in the world are kept in some sort of stall for at least part of the day/night cycle, if not all of it. Often within a closed structure, like a barn. Some stalls are bigger than others, but the vast majority of box stalls in closed structures are approximately twelve by twelve feet. The accumulation of negatives from this lifestyle is devastating to an animal born to be outside, on the move, with the herd, day and night.

The most frequent argument we've heard is, This isn't a wild mustang, it's a domesticated horse. As if the declaration "He isn't running free" would somehow change the millions of years of genetics that have made him what he is. As if such a statement would make the ammonia from poop and pee eating away at his feet disappear; or cause his physical structure, which was built to be on the move constantly, to be suddenly fine with standing still twelve to twenty-four hours a day. As if it would make his respiratory system, which is built to be outside breathing fresh, clean

air, suddenly find it healthy to breathe in ammonia and high
quantities of carbon dioxide in a closed environment. The average horse breathes 62 liters of air a minute, producing 150 liters
of CO_2 per hour. And ammonia is so destructive to protein, it is
actually being taken off the market in some countries.

Not being a wild mustang does not compensate for the reduced
blood circulation a horse wearing metal shoes suffers while
standing still in a stall. Reduced circulation in turn weakens the
hoof by reducing the quality and quantity of horn produced by
the hoof. And reduced circulation that doesn't efficiently pump
blood back up the legs to the rest of the body adds stress to the
heart and affects the immune system.

And whether mustang or domestic, it isn't healthy for a horse
to eat from a bucket, feeder, or hay net usually hanging at table
height when his body is built to eat from hoof level. Nor does
being domestic negate the claustrophobia and stress he lives with
on some level, caused by feeling trapped, unable to flee, alone,
and bored. Never mind how willing he might be to go into the
stall either because he has always been forced to or because he
knows that is where the food is.

Is it any wonder that domestic horses, on average, do not have
near the life span of horses in the wild?

The wonder is how so many caring, intelligent, conscientious
people have remained so uninformed about what they're doing to
their horses. This information is readily available. In studies. In
books. On the Internet. Backed up. In depth. With consensus.

The wild horse model works. It's simple to create. And the
horses are not only healthier, they're happier.

Just ask Scribbles. Or Cash. Or Mariah. Or Pocket. Or Handsome. Or Skeeter.

9

Bloodlines

The golden stallion circled the herd, watching, waiting. It was his job to protect, but he was always keenly interested when one of his loin was to be born. In the center of the herd, the matriarch lay on her side, new birth imminent. Unlike many stallions, the palomino felt attachment. A new foal was noble, and he treated it so. Perhaps the quality of his lineage, and all that his ancestors had been through, had somehow found its way into genetic code. Or maybe it had always been there. The blood of a stallion who had traveled the entire breadth of this great land ran through his veins. And that of one who had saved his herd from shipwreck. His pride was justified. And, once again, he was passing it on.

THE STORM HAD hit the small fleet the day before at dusk, unexpectedly and with vicious intensity. The ships were immediately scattered and lost contact with each other.

Swinging helplessly in his harness on the smallest of the five ships, the big stallion expected to be ripped apart at any moment by the driving forces of the crashing, breaking waves. He tried to twist around to check on his herd, but each wave came larger and more menacing than the last, smashing over the bow, sending torrential rages of angry sea across the decks. Were his mares still alive? He had no way of knowing.

Twenty-six humans were belowdecks, crammed into a space barely large enough for ten, and they were all violently ill. If not from the turbulent sea, then from the vomit of those less sturdy. Fifteen horses were swinging like wind chimes from the rigging of the ship. One mare had slammed into the gunwale and fractured a leg. It was miraculous that, so far, hers was the only severe injury. But the palomino stallion knew none of that.

Another explosive onslaught of foaming seawater drove his feet off the deck and slammed him toward the rail, testing the strength of his harness. He managed to lift his legs and swing clear of the gunwale. He knew he was going to die but not without a fight. He snatched at one of the rigging ropes supporting him. It would be better to be in the sea.

Across the big ocean, he had often swum in the sea before he was captured. He and his entire herd. It was a sad day when they had been caught off guard. A day just like this one, with no expectation that humans would be about. The beach, with waves crashing and winds blowing, had seemed a safe place. But when the humans appeared, there had only been two options. Swim into a sea as stormy as this one or run for a small canyon and hope for the best. The matriarch had taken her only real choice, but the best didn't happen.

The stallion was descended from proud Arabian stock brought to Spain after the Crusades, and his mares were mixed with his blood and that of the Barb. And now, because he had not protected them well, they would all surely die.

Suddenly he felt the small ship lurch sideways and lumber into the trough between two huge swells. He struggled to look toward the helmsman, but no one was there! No one was steering the ship! He felt himself rise and he knew what was coming. The boat was drawn upward into the vicious curl of a breaker three times its size.

There was a loud slap and the stallion seemed to fly through the air. Then a crack and everything was black. Black, and cold, and wet. He suddenly realized that his legs were churning and he was free of his harness. He held his breath as the raging water swirled around him. And he churned his legs, reaching upward, reaching for air.

That he knew which way was up was a mystery. But that he was free was a miracle. The ship had smashed onto a shoal, then been lifted and flipped like a leaf in the wind, turning totally over; the loose ballast, stores and cargo that normally lay in the bilge and kept the boat upright, had crashed through the cabin sole and would keep the little ship upside down.

The big stallion's head bobbed above the crest of a breaking wave. Suddenly there was air! Air to breathe! A huge wave of cold, green water slapped him in the face, but he didn't care. He had air and he could swim. And there was land! He could smell it. And he could tell it wasn't very far away.

That night thirteen horses slept on the beach of what would one day be called Shackleford Banks, North Carolina. They slept longer and deeper than horses usually sleep, and before dawn a large palomino stallion had them moving across the dunes into a stand of trees. Before the day was out, they found one additional horse from the ship, leaving only one who had not made it. The matriarch. Her shattered leg had left her unable to swim. In time she would be replaced. But for now, all had eaten heartily of the abundant marsh grass and had traveled the length of the seven-mile island. They even ran for a brief spell, kicking up their heels and tossing their heads. They had no idea where they were, but they knew they were alone.

And safe.

With food.

And the stallion was pleased.

10

Survival

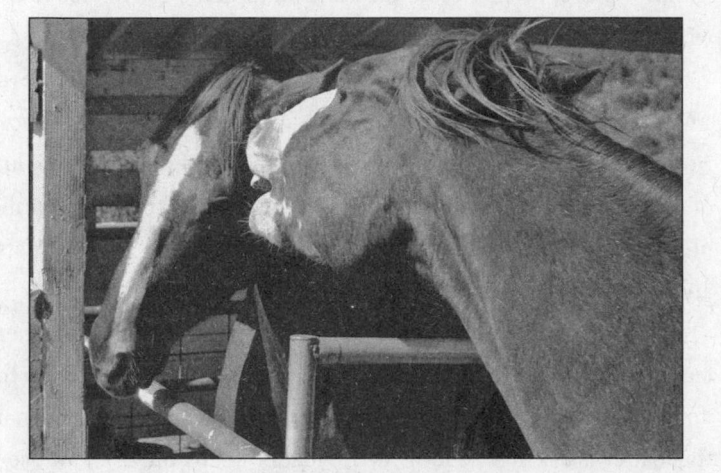

Survival.

This is the number one instinct of the horse.

To survive, the horse needs food, water, and safety. That's it. Along with procreating, that's pretty much all they think about every waking hour.

Much of the survival drive is wrapped up in the instinct to be safe, which means being part of the herd, understanding the language of the herd, and understanding the social order of the herd. Every herd, no matter how large or small, has a distinct pecking order. All determined by who moves who, thus who respects who, which translates into who feels safe with who as their leader.

So where does that leave *us*, the humans in their lives? Many people have told me that if they let their horse live with a herd 24/7, they would lose their relationship with the horse. The horse would forget all about them. Ignore them.

This is simply not true, *if* the relationship begins correctly in the first place. *If* the horse was allowed to make the choice to be with his human. *If* the human has proven to be a good leader.

The object for the human, then, is to become part of the herd in a leadership role. And no matter how it has been made to look in movies and television, leadership of the herd is not forged out of fear of the strongest horse. It is much deeper and more complex than that and has a great deal to do with who the herd trusts and respects. And who they believe will keep them safe.

Likewise, when a horse has chosen a human to be his or her leader, along with that choice comes an implied responsibility to do what the leader asks. So long as the horse understands what the leader is asking. So long as the leader keeps the trust and respect of the horse. And so long as the horse feels he or she is safe with this leader.

Such a relationship does not dissipate just because the horse spends time with other horses. The human is now part of the social order. When Cash leaves the pasture, whether for hours, days, or weeks, he is still the leader of the three in his clan when he returns. He knows it, and the other horses know it.

And I can walk in at any time and have a special moment with him, or with any of the other horses. Even with halter and lead rope in hand. No one runs or hides. Often they'll come to me and follow me around. I'm not just one of the guys, I'm one of the *respected, trusted* guys. One of the *safe* guys. As is Kathleen. They know that whatever comes, it's not going to be bad, and will most likely be good. And we make sure that's the way it works out.

The leadership must be genuine, which for us has meant spending time with the herd absorbing the way they make decisions. How they discipline each other. And how all that translates into respect, and trust. For a while, this was difficult for Kathleen. As she worked her way through her fear, she would often try to fake it by yelling and waving her arms at the horses to assert herself, which she didn't realize was really a step toward dominance. And it scared the horses.

"A leader doesn't act like a wild person," I finally told her. "All you need do is swell up like a horse and pin your ears."

"I can't pin my ears," she grumbled.

"Yes, you can," I said. "Figuratively, you can. Watch."

I turned my back on Scribbles, who was always trying to invade our space, get close, nibble at my hat or shirt. And, sure enough, here he came. I spun around, looked him straight in the eye, and swelled up like a hot-air balloon. A flick of my finger and the words *back up*, said firmly but quietly, were all it took. He stopped in his tracks, and took two steps backward. I rubbed him on the forehead and scratched his ear.

"Good boy."

The next morning at feeding time, Kathleen worked the same magic on Mariah, who was always trying to steal Skeeter's feed from her. And she amazed even herself. That afternoon, when I couldn't find her, I wandered over to the pasture and there she was, in class. Soaking up the way the horses do it.

Respect does not come from bossmanship. And conversely it is not given to someone who showers horses with baskets full of love, without discipline. Sit and watch a herd sometime. Just watch. Make note of who respects whom, and how it's shown. *I respect you enough, and trust you enough, to subordinate my judgment and safety to you.*

Pretty powerful when you know that, to the horse, safety means survival.

And, again, for us humans it means "love, language, and leadership in equal doses."

That *equal doses* part is what kills most relationships.

Does the person who can get her horse to come only because she's offering treats gain the horse's respect? No. That's just telling the horse that her presence means food. You want your horse to come because he or she wants to be with you. And when you begin by giving the horse the choice to be with you, and when you learn to communicate from the horse's end of the lead rope, creating that willing relationship is totally doable. It is never too late to begin again.

Does the person who lavishes love on a horse without discipline and training garner the horse's respect? Watch the antics of the herd. The respected horses are the leaders. Those not respected are being led.

Does a person get respect by charging into a stall or a pasture, immediately haltering her horse, and dragging him straightaway off to work? Without, maybe, pausing to speak with the horse, or rub him, or just hang out for a bit? Perhaps asking for permission to halter?

Sound silly?

Try it. I hang out, at least attempting to put the relationship before my selfish desire to go ride or train. I rub and talk a bit. Sniff some nose. And almost never put a halter on a horse who hasn't offered to help. It makes such a difference. I believe if you give respect, you'll get it back tenfold.

Don't get me wrong. I don't treat horses like puppies. I treat them like partners. Junior partners, of whom I expect great things. And so far they've all delivered more than I've asked. Because, for the horse, to acknowledge and respect a leader is to feel safe. This is deeply rooted in their nature. And feeling safe means survival. Which makes the leader the source of emotional comfort.

Is it any wonder, then, that they work harder for a good leader? Don't we all?

11

Relationship

It had been a long and hot day for the young colt, made hotter and longer by the weight on his back. But he wasn't complaining. They had found each other, the colt and his boy, almost a year ago on the island where the colt's ancestors had lived for ten generations. The colt had been only a foal when they met, but a proud foal. A direct descendent of the mighty stallion who had endured the shipwreck and led his herd onto the island and taught them to survive.

Now the colt and the boy would have to do the same. They were both without homes, without family. The boy was a young Powhatan, one of the few remaining from the great chief's band of tribes scattered through Virginia and the Carolinas. After his parents both died of English disease, his grandmother raised him until she, too, died. The boy had seen the island horses once on an outing with his father and had decided then that one day he would be family with them.

The colt had seen some cruel humans during his short life, but the boy was different. When he first appeared on the island, the boy spent many days camping well off away from the herd, watching them and studying their communication. And showing them he was no predator. When he moved closer, he would sit for long days on a fallen tree trunk, only watching, making no attempt to confront or pursue. When the sun disappeared across the sound, and daylight turned into darkness, he would sleep under the fallen tree, wrapped in a thick blanket his father had traded for with an English man. The boy carried no weapons, no ropes. Just the blanket.

Gradually, the herd became curious and wandered closer, encouraged by the kindness in the young boy's eyes. One cold winter morning, huddled in his blanket under the log, the boy awoke to the warm breath of horse. He slowly opened his eyes to see the nostrils of the foal, sniffing, puffing. He did not look the young horse in the eye as a predator would, but stayed focused on his nose, and he, too, puffed and sniffed a greeting. The foal stepped back, and the boy sat up, wrapping the blanket around his shoulders. He turned his back to the young horse, his head down, his shoulders slumped in a show of friendly submission, saying, I am approachable, I am not a predator. He had learned this from observing the herd. It was the next day before the foal actually touched him, and only then did the boy reach out and rub the young horse on the nose, then on the forehead.

Since that day, they had become very close. When the time came to be shunned from the herd or fight the stallion for dominance, the young colt chose the boy. They swam together across the sound to the mainland and were now far away, traveling through the wilderness more or less following a group of boats on what the boy called a river. The young Powhatan was excited about seeing new lands.

The colt didn't understand it all but was having a wonderful time. They would run like the wind through the trees and on the riverbanks, with nothing between the boy and the horse but the boy's leather breechcloth. The colt could feel the boy's every movement, every pressure, and the two had developed a language of what each movement and pressure meant. The colt and his boy were like one.

At the moment, they were standing on a tall bluff, looking down on the boats traveling upriver. They could see their friend, a man as black as the colt's mother. He wasn't hard to pick out of the group of so many white men. The boy had called him York when the man had arranged for them to cross a big river on one of the boats. Along the journey, the boy and the colt had been helping the man hunt for food for his master, one of the leaders of the white men, and York had appreciated their help and was anxious to keep them tagging along.

The boy swung himself up onto the colt's back, clung tightly to his mane, touched his neck, and nudged him with his calf. Imperceptible requests to anyone who might be watching, but the blond colt knew exactly what to do. He spun on his hindquarters and trotted off through the woods. The colt could actually feel the thoughts of the young boy. It was a good partnership.

12

Connection

T he sun was low in the sky and the cool breeze from the ocean was at last making its way through the mountains and overcoming the heat of the day. Usually I would give Cash a bit of a massage, brush him down, and clean his feet at the very least before saddling up, but on this day I was thinking about the young Indian boy and the colt. The breeze felt good and I felt spontaneous. And I wanted to be close to my horse.

I looped the lead line on his rope halter, like reins but with no bridle or bit, dragged out the mounting block, and climbed aboard his bare back, my creaky bones well beyond the days of merely "swinging up." I settled in and he felt good. His muscles twitched under my rear and he glanced over his shoulder and gave me a funny look. He's always giving me funny looks; some I can decipher, some not. He's the most expressive horse I've ever met.

I touched just the legs of my pants to his side. Just a touch, and he walked off.

About halfway down to our little arena I realized I had never actually ridden down. And I remembered why. The trail is *very* steep. I had ridden up several times, but never down. Now I knew what the funny look was about when I climbed on. *Whoa! You're out of pattern. Are you sure you want to do this?*

I was slipping and sliding toward his withers—*ouch!*—pushing off against his neck. He paused to make sure I was okay. Then we went on.

This is a horse who is very impulsive. I suppose the term *very* is relative to one's experience level, but let's just say he does like to go and, as good as he is at most things, he has yet to completely comprehend the maintenance of his speed over long periods of time. But whenever we're bareback, he is clearly aware that I'm more vulnerable and he always minds his p's and q's. For one reason only. Because he chooses to.

He felt good under me. I was thinking about the young colt and his Powhatan friend, and wished I was a better rider. I would have loved to race the wind bareback on this horse, like the young boy.

During the year and a half since Kathleen and I had leaped so innocently into the world of horses, I had worked with Cash to develop a relationship and communication like that of the colt and the young boy in the story. *What does this touch mean? And that one? And does this work better than that?* But the saddle always seemed to be a deterrent to this new form of communication. It got in the way. When I was in the saddle, we'd both try, but progress was slow. I was never sure if I was in the right spot or if I had applied enough pressure. I couldn't feel him. Couldn't sense his feel of me.

That's what spurs are for, one crusty old wrangler told me. You can probably guess my response to that.

Then one day I decided to have a go at it without the saddle. Cash was doing so well on the ground, I decided to not only ride bareback, but to use just a halter and a lead rope. No bridle and bit. And that was the day.

It was exhilarating. The difference was so extreme. It was like feeling the ocean's currents in a bathing suit versus sitting in a deck chair on a cruise ship. We developed our communication and learned well together, and when I returned to the saddle, we both knew what to do. I never enjoy riding with the saddle as much as bareback, but now we both understand the communication. I recommend riding bareback to any novice. Not on just *any* horse, of course. Be certain on that score. Don't put yourself in harm's way. Do your homework. Be sure you have a calm horse, with good controls, and a relationship you have nurtured. Then it can be a life-changing experience.

I enjoyed our ride immensely that day, but it was what happened *after* the ride that caused me to rewrite this chapter and wipe tears from my eyes.

Cash and I were trotting back up the hill from the arena, me clinging to his mane to keep from sliding off his butt. Kathleen was making her way toward the tack room, where we often spend a few moments in the late afternoon relaxing and talking about our day. We have the only tack room I know of with a back deck, overlooking our little arena and the setting sun.

But on this particular afternoon, we didn't venture out on the deck. We just plopped down on the tack room step to enjoy each other's company. Kathleen, me, and Cash. After a bit, I decided to remove his halter, thinking he would wander up to the little patch of grass, maybe thirty feet away, where it would be fine for him to nibble on a snack before dinner. I asked him to lower his head—it's not a very tall step we were sitting on—and I untied the halter. But Cash didn't leave. He nuzzled my nose, sniffed, checked out our water bottles, nibbled on my hat . . . and just hung out. For maybe fifteen or twenty minutes! It made my night, and the next day, and here I am still talking about it. How good it feels to know that this horse whom I love so dearly really likes being with me. With *us*, a part of the family. Part of the herd.

Relationship.

Choice.

Connection.

13

Off with the Shoes

It's been a year and a half since Cash eased up behind me in that round pen and nuzzled my shoulder. It changed my life, and I hope his. It made me dig ever deeper to discover the right answers for this horse, and to not just take someone else's word that this way or that way is the right way.

From the very beginning I was worried about his feet. Ironically, not because he had metal shoes nailed to them. But because he only had *two* metal shoes nailed to them. On his front feet. His rear feet were barefoot. It's a quarter of a mile walk on hard asphalt from our place to the local horse club arena. I was worried about his back feet having no shoes.

"If his front feet need shoes, why don't his back feet?" I asked everyone.

"Because 60 percent of a horse's weight is on his front feet" was a standard answer. In the beginning I was too intimidated to ask why a mere 20 percent meant the difference between shoes or no shoes. I still haven't heard a good answer to that one.

Here's what I was led to believe a mere year and a half ago: Bare hooves banging against hard surfaces, be they concrete, asphalt, dirt, or rocks, will cause the hoof walls to crack, and shatter, and crumble. Therefore horses need metal shoes.

At this point I didn't know that such cracking, shattering, and crumbling doesn't happen to the hooves of horses in the wild. Their hoof walls and soles are like steel. And there's a reason for it. But nobody ever told me that. And nobody told me that a metal shoe is so unhealthy for the horse's hoof that it can become the cause of cracking, shattering, and crumbling. I began this leap into the world of horses just like everyone else, because everyone else was who I was listening to.

It was the relationship with Cash that made me keep digging, keep reading, and keep learning. Because I cared deeply for this horse who had chosen me, who had handed over to me so much of his livelihood, I was left with no choice but to do as right by him as I possibly could. And that meant I needed to gain knowledge.

Still, it's disturbing that once upon a not-so-distant time, I firmly believed that horses *must* be shod. That's the way it was. Horses wore shoes. It never occurred to me to find out why horses in the wild, or in the past, got by without nails and shoes. And that's worrisome. I consider myself a reasonably intelligent and curious individual. Why would I accept something so odd without even asking a question?

Scary.

But common, I've found, among so many horse owners.

I was a novice blithely following whoever would speak up.

In his seminars on leadership, Andy Andrews tells folks that the first step in becoming a leader is to be a person of action. When everyone else is shrugging their shoulders, *do* something. *Say* something.

Hey, where do y'all want to go eat tonight?

Gee, I don't know. Where would you like to eat?

Makes no difference to me. How about you, John?

Oh, I'm good with anything. Bill, you decide.

Oh, I don't care. Really.

"Just make a decision! Be a person of action! Name a place, any place," screams Andy, "and suddenly you're a leader!"

Let's go to McDonald's!

Hey, good idea.

Fine with me. How about you, Bill?

Yeah. McDonald's. That's good.

Take action. Step out. Speak up. And they will follow.

Do it two or three times and they'll be looking to you whenever a decision needs to be made.

Never mind whether or not you make good decisions. Or healthy ones. Or even whether you make any sense. Just taking action puts you at the front of the line. And, unfortunately, that's how so much *mis*information gets spread. Somebody somewhere has the chutzpah to say something. That person might or might not have knowledge on the subject. He might or might not have ulterior motives. But he spoke out when no one else would. When no one else wanted to take the time to think about it, or research it. And now, voilà, he's the expert.

That's the way it was with our horses and shoes. Everybody said do it and I didn't question the advice. They had to know more than I did, right? After all, they had been at it for years. Some of them for decades. Who was I to question?

Then I stumbled upon an article in a horse magazine. The first couple of paragraphs went something like this:

Did you know that a horse's hoof is supposed to flex with every step taken? And that simple act of flexing is just about the most important thing a horse can do for good health and long life? The flexing provides shock absorption for the joints, tendons and ligaments in the leg and shoulder, acts as a circulatory pump for blood in the hoof mechanism, and helps the heart get that blood flowing back up the leg.

Without flexing, the hoof mechanism will not have good circulation and will not be healthy. And the heart will have to work harder to get the blood back up the legs. Without flexing, there will be no shock absorption.

And with a metal shoe nailed to the hoof, no flexing can occur.

Kerwhap!

Slapped right in the face with a piece of indisputable logic.

Of all things!

Logic!

Truth!

How insensitive to my inertia.

The article went on to explore the results of a study of more than a thousand wild mustang hooves. All barefoot hooves, of course. All very much alike, healthy and hard as steel, without re-gard for the type of home terrain or climate.

I was immediately off to websites, gathering books, soaking up more information and knowledge.

Pete Ramey, a world-renowned natural hoof specialist in Georgia using the "wild horse trim," says he has never worked on a horse that he has not taken barefoot *successfully*.

Eddie Drabek, a specialist in Houston also using the wild horse model, helped take the entire Houston Mounted Patrol barefoot, and they are all doing terrifically, with lower vet bills than ever before, even though they work every day on concrete and asphalt. When they were wearing metal shoes, the horses were often ice-skating on the asphalt, and horses would go down with their riders. But not one has fallen since the barefoot pro-gram began.

Drabek has also taken winning performance horses barefoot in reining, jumping, racing, cutting, and dressage, making for happier, healthier winners. He says, "Every single horse brought to me that the owners swore had feet that grew abnormally, had bad genetics, could never be barefoot, had brittle hooves, had cracks that would never go away, and so on and so forth—I've heard it all—has been taken barefoot successfully with the proper balanced trim and has beautiful feet to show for it. And I'm talk-ing hundreds of horses."

Ramey, Drabek, and natural hoof specialists Jaime Jackson, James Welz, and Marci Lambert have all taken on one or more horses that a vet or a farrier has said must be put down because of lameness in the feet, and in every one of those cases, the horse, after going barefoot with the wild horse trim, completely recovered and became perfectly healthy. Some healed quickly, some took as long as a year, but recover they did. Case history after case history.

So why do people believe their horses cannot exist without shoes?

Because, as Dr. Strasser, Ramey, Drabek, Welz, and Lambert all confirm, when a shoe falls off a horse that has been shod for years and years, the hoof and hoof wall are usually no longer strong and healthy. The hoof has been made unhealthy by lack of circulation because it has not been able to flex and thus circulate the blood properly throughout the hoof mechanism. And the continuing process of hammering nails into the hoof wall makes it weaker, and provides places (the nail holes) for chips and cracks to occur. Also, some hooves, if they're in really bad shape, will be tender for a while after going barefoot. And the unknowing owner concludes that the tenderness means the horse *needs* shoes.

Not so.

The hoof will completely heal and remodel itself, growing a strong new horn and a hard calloused sole. This is a logical and normal process (see the Resources section at the back of the book, especially Pete Ramey's and Jaime Jackson's books and videos). It takes approximately eight months for a horse to grow a new hoof, from his hairline to the ground. If properly trimmed to mimic the way wild horses' hooves trim themselves during daily wear, a worst-case scenario for a horse to acquire a completely new, rock-solid, healthy foot, then, is approximately eight months. Many horses are much quicker. As you read earlier, Cash was good to go from the first day his shoes came off. And a happy horse indeed. Four of our six never had a tender moment after going barefoot. One took four months to remodel, and one took almost seven months. And well worth the time.

But wait! When my horse's shoe falls off, he starts limping almost immediately. And when the shoe is nailed back on, suddenly he's fine. Doesn't hurt anymore. Proof that the shoe is better for him than barefoot.

Have you ever crossed your legs for so long that your foot goes to sleep? We all have, and we all know what's happening. The leg-cross has cut off the circulation to the foot, and with no circulation, the nerve endings lose their sensitivity and fail to work. The second you uncross, or stand up, the circulation returns, as do the nerve endings.

Ooh! Ouch!

The same thing happens to a horse when a metal shoe is nailed on. The inability of the hoof to flex removes its ability to pump blood, virtually eliminating circulation in the hoof mechanism. Without proper circulation, the nerve endings quit transmitting, and the horse no longer feels the "ouch." When the shoe falls off, the circulation returns and suddenly he can feel again.

Whoa, what's that about?

As mentioned earlier, Scribbles took a good six to seven months to regain a healthy hoof with no ouch. And today he's a happy camper, on asphalt or concrete, on the trail, in the arena. His hooves are beautifully concaved, keeping the coffin bone up where it belongs. They are beveled at the edges, just like a wild horse's hoof. And they're as hard as stone.

The sacrifice? The downside?

A few months' time to let him grow the hoof nature always intended for him to have. Good boy, Scribbles.

How much trimming is needed, and how often, depends upon

the horse's lifestyle. Remember, the objective is to replicate the hoof that the horse would have if he were living in the wild, moving twelve to fifteen miles a day with the herd. If he's living in a box stall and not moving around much, there will be a lot more trimming, probably more often, than if he were living in a natural pasture like ours and moving around all day wearing down his own hooves. But the objective can be reached in either case.

Pete Ramey, the hoof specialist mentioned earlier who teaches hoof care all over the world, believes we have only just begun to discover the true potential of the wild horse model. After a trip to wild horse country for research, he said, "The country was solid rock; mostly baseball-sized porous *volcanic* rock that you could literally use as a rasp to work a hoof if you wanted to. Horse tracks were fairly rare, because there was so little dirt between the rocks. There were a few muddy areas from the recent snowmelt, but they were littered with rocks as well. The horses made no attempt to find these softer spots to walk on."

Pete and his wife, Ivy, observed, videotaped, and photographed at least sixty horses. All of them, from the foals to the aged, moved effortlessly and efficiently across the unbelievably harsh terrain. According to Pete, the horses were doing extended trots across an obstacle course that would shame the best show-ring work of any dressage horse, with their tails high in the air and their heads cocked over their shoulders watching the intruding couple.

"I have never known a horse I would attempt to ride in this terrain," Pete says. "Ivy and I had to literally watch our every step when we were walking. The movement of the horses was not affected by the slippery dusting of snow on the rocks. In fact, they got around much better than the mule deer and the pronghorns.

The entire time we were there we did not see a limp, or even a 'give' to any rock, or a single lame horse, and not one chip or split in any of their hooves. It was an unbelievable sight."

The world has been shocked and amazed by the ability of Pete and others to forge rock-crushing bare hooves, boost equine performance, and treat "incurable" hoof disease. "I don't want to diminish these facts," Pete says, "but I now realize that we haven't even scratched the tip of the iceberg."

There's an old expression: *No hoof, no horse.* And the reams of research I've pored over truly made that point. So much of what can go wrong with a horse begins or is controlled by the health of the hoof. When that hoof is healthy, flexing, and taking stress off the heart, it can add years to a horse's life.

And he'll be happier.

Not only because he feels better, but because he can actually *feel* the surface he's walking on, which makes him more comfortable and more secure in his footing. It's the way nature intended. Would you run on the beach with boots on? Or do you want to feel the sand between your toes? Not a perfect analogy, but you get the idea.

The president of the American Farrier's Association, in a speech to his constituency reported in the organization's publication, said that 90 percent of all the domestic horses on this planet have some degree of lameness. Dr. Jay Kirkpatrick, director of the Science and Conservation Center in Billings, Montana, has studied wild horses most of his adult life and says that lameness in the wild is extremely rare and virtually every case he's seen is related to arthritic shoulder joints, not hoof problems.

Off came our horses' shoes.

Once onto wild horse model, I was in the soup, so to speak. I began to question all sorts of other things. Like blankets, leg wraps, barns, stalls, feed, the horse's nature—and what I found was like a jigsaw puzzle. Start goofing around with one piece and it affects the whole picture. And the picture I was beginning to see more and more clearly was that we humans have completely manipulated horse care and training to suit ourselves, not the horse. Under what kind of leadership did we ever get to this place in time? Why has all this information not been front and center? I posed these questions to Dr. Matt, our vet, and began to think I was speaking to a politician up for election.

"Well, it's not always so black and white" was his answer to a question about going barefoot. "I like to see horses barefoot whenever they can be."

"Why not always?"

"Well, some horses have issues that others don't have."

"Like what, *owners?*"

The smile that wiggled across his lips betrayed his words.

"Well, some people feel their horses need shoes if they're going to be jumping, or, say, doing endurance riding."

"What do *you* think?" I asked.

"I think some horses have issues other horses don't have."

I was getting nowhere, and he had to keep moving. Another client to see. I knew from my own research and experience that he was an excellent vet. A caring vet who owned horses himself and loved horses. He had good communication with them. So I couldn't figure out why he was avoiding my questions. Was I wrong? Were there circumstances I hadn't yet stumbled upon in my digging that would refute the entire wild horse model? Was he simply not well informed on the work of this ever-growing band of natural trimmers?

I called the next day and asked if I could take him to lunch.

And told him why.

"I need to hear answers from you," I said. "I want to know what you really believe. I'll promise to never repeat what you say, if that's what you want, but I need to know that I'm not crazy. Everything I've been studying says that most of us are doing virtually nothing right in the way we care for our horses." What he told me the next day chilled my blood and made me very sad. And I'm afraid it's only a microcosm of the way too much of our world works today.

A farrier is a person who makes a living putting metal shoes onto horses' natural feet by nailing into the horse's hoof wall. He used to be called a blacksmith. One organization reports that there are probably 100,000 farriers on the planet. The farrier's livelihood and self-esteem are generated by how well he appears to do his job. How well the shoe fits. How well it seems to solve some problem with the horse's foot, like an imbalance. Or an ouch. He decides whether the hoof needs a pad, or some packing, or wedges, or a special type of shoe. Often the farrier does no hoof trimming. His assistant does that. The farrier shapes the metal shoes and nails them on. A natural trimmer wrote about how difficult it was to stop shoeing and *just* be a trimmer using the wild horse model. He said it was an ego thing because his assistant usually did the trimming, and now he was more or less doing his assistant's work himself. He also called it a "male" thing, because he really liked the molding and shaping of steel and hammering nails. I spoke with a couple farriers about switching to the concept of trimming barefoot horses with the wild horse trim.

"It's bullshit," said one. "A horse needs shoes."

"Why?"

"Because we've bred the good foot right outta them."

"Have you ever tried it? The wild horse trim, I mean."

"Nope. Don't ever plan to."

"I know of several natural trimmers who have never been un-successful taking a horse barefoot."

"Anybody can take a horse barefoot. Just pull the shoes."

"I mean successful in that the horse never needed shoes after-ward. Had a healthy, happy foot. On the trail. On the road. In the ring. Wherever."

"Bullshit," he said, which pretty much ended the conversation. Not exactly my kind of logic.

There are a lot of horses in our community, and therefore a lot of farriers.

Dr. Matt told me how things are out in the field. He gets to see a client and treat a horse, usually, only when there's something wrong. An illness or an injury. In other words, rarely. A farrier usually sees a client every six to eight weeks, maybe eight to twelve times a year. So most horse owners know their farrier much better than they know their vet. If it's a long-term rela-tionship with the farrier, it would stand to reason that he is trusted. One bad word from the farrier about a particular vet, or a good word about some other vet, will be heard. And a farrier is not likely to recommend a vet who he knows is going to come in behind him and tell the owner to pull all the shoes off his horses.

Even with existing clients with whom he has had good rela-tionships, Dr. Matt has lost patients because he recommended that shoes come off.

The owner calls the farrier about pulling the shoes.

The farrier explains that "most vets don't know much about feet because they don't work with feet. And, well, you should really think about it before pulling the shoes." Those words were actually spoken to me by a farrier.

In the above example, either the vet or the farrier is usually going to wind up losing a client because the last thing owners want are folks who disagree about the treatment of their horses. Especially if the owner doesn't have a clue about which one is right.

The very sad thing about all this is that all the equine vets in the country should be educating themselves on the magical things that can be accomplished with the barefoot wild horse model. And they should be talking to clients about it. But the truth is that it would be difficult indeed for a vet to make it in a community in which he has alienated all the farriers.

There's a vet in a neighboring community who actually stocks horseshoes and farrier tools and sells them to farriers at a discount. I'm guessing that doing so wins him a basket load of recommendations from farriers. Is he likely to tell his clients to pull off the shoes he himself sold to the farrier who nailed them on?

There is no disputing that a horseshoe prevents hoof flexing. Nor is there any dispute about why the hoof is supposed to flex. Nor about the good things that happen when it does. I can't help but wonder how a vet sworn to do his medical best for horses can sell horseshoes and supplies to farriers and still live with himself.

But I didn't press, and changed the subject with Dr. Matt.

"I've read that leg wraps are not good for horses," I said. "The article stated that they're usually tightly wrapped when the horse is at rest; then he goes out to work and the blood vessels in the leg attempt to dilate to get more blood down to the working leg, and the leg wrap prevents the vessels from dilating. True or false?"

Dr. Matt smiled. "I don't think they give any support or any true protection for the leg, but if they aren't worn too tightly, they don't really hurt anything."

"But isn't it really best not to have them at all?" I persisted.

"Look at it this way: If an owner wants to use them, and I tell him no, and the horse comes up lame from some activity, who's going to be blamed?"

"Blankets?" I questioned, again changing the subject.

"No need for them unless it's really cold and raining. A horse has a terrific system for keeping his body temperature where it needs to be, unless his coat gets really soaked while it's really cold. Snow is no problem. It's the combination of cold and wet. I recommend pulling them as soon as the rain stops to keep the blanket from weakening the horse's own internal system."

"Is there any hard research on cold and wet?" I asked.

"Better safe than sorry," he said.

"So, most of the time, the owner is blanketing his horse to make himself feel better. Like I almost did."

"I prefer to think you were more misinformed than selfish." He smiled.

On the subject of stalls and barns, he did say that horses are better off moving around, being out 24/7. I was jubilant. At last an unqualified recommendation.

"Do you recommend that to your clients?" I asked.

"I try to be sensitive. If a person can do nothing but provide a twelve-by-twelve stall, there's little point in telling him to do something different."

I believe if a person loves his horse, he'll figure out a way to do what's best for him. Or at least put some thought into it. But, again, I didn't press.

"There's an old saying," Dr. Matt said. " 'You can have money. Or you can have horses. But you can't have both.' I usually get called as the last resort because people don't want to spend money for a vet. Remember that call I got yesterday while I was at your place? When I got over there, I was told the horse had been lying on his side without eating, pooping, or peeing for three days!

"*Three* days!" he added incredulously. "I deal with that kind of thing every day of my life."

"There are people who shouldn't have horses," I said.

He nodded.

I quietly thanked God for Dr. Matt, because I couldn't do what he does. I couldn't face what he faces every day. I'd have no clients at all by the end of the first week.

He did share that he felt there was a new day on the horizon. "For several generations, the horse was nothing more than a beast of burden, like an ox. Or a tractor."

"Or a motorcycle," I said.

"Right. But today I'm seeing more and more people who actually care for their horses. Granted, it's a small number, but it's growing. With all the publicity that people like you are getting, and the natural horsemanship clinicians, and the barefoot trimmers, and the vets who have studied all this . . . well, I have to believe it's getting better."

"I hope so," I said.

But, unfortunately, for the most part, word won't be coming from the folks whom you would normally turn to for advice. The farriers, most of them, are not going to take up the banner of barefoot. If they did they would have to completely change their skill set or they'd be out of work. A few of them have done just that, but *few* is the operative word. We've already talked about how difficult it is to get most people to change, even when it's change for the better.

The veterinarians, most of them, have no choice but to hedge for the same reasons. Economics. Fear of being out of a job. Fear of risk. I've spoken to vets who have said, "I agree totally with what you're saying, but please don't tell anyone I said so." Many of them feel they can serve best by keeping their jobs and making a few small inroads here and there. Considering what they face, it's difficult to argue with that.

Until I look into my horse's eyes.

Letting him be a horse certainly won't be getting endorsements from the manufacturers of horseshoes, leg wraps, blankets, prefab barns, hay feeders, and so on. Those folks aren't going to burn their paychecks.

So think about it. Think seriously about it every time you hear someone say that what they do for a living is better for your horse than what the horse would do for itself in the wild. Ponder the presidents of those tobacco companies testifying before Congress, emphatic that tobacco was not harmful. Dig around on your own. Do some research. Compare what "the experts" say. Gather your own knowledge and don't let someone else make decisions for you. Whether it's about your horses or your life.

And if you do own a horse, show him that he's not an ox, or a tractor, or a motorcycle. He's your partner.

And let him know by your leadership that you love him and will give him the best care you possibly can.

14

Nature Lives

For more than a year, the golden colt and the young Powhatan had been following the band of river travelers heading ever toward the setting sun. They had seen and experienced things that were beyond description. Great buffalo in herds of thousands. Eagles, hawks, trees taller than mountains, and mountains taller than the sky itself. Indian warriors who, like the boy, raced the wind on the backs of horses.

And snow.

This unusual white substance that came from the sky was of no consequence to the colt, other than it made him feel playful. But the cold that came with it seemed to take a severe toll on his young friend. The boy had traded a deer felled with his bow to a young Shoshone about his own age, receiving in return a cloak made of thick buffalo hide. This seemed to keep him warm most of the winter, except for those days and nights the colt had to lie across a hole in the ground to protect the boy from the snow and icy wind.

The young stallion's coat had grown thick and long, and the boy once joked that he looked more like a fluffy bear than a horse. On these cold days of winter, if the colt sensed rain in the air, instead of snow, he would lead the boy to a windbreak—a cliff, an outcropping, or a stand of trees—to avoid the combination of extreme cold and wet that might penetrate his coat. The winters in this new country were harsher than those in his native North Carolina, but he adapted readily, as his ancestors had done forever.

The boy had watched the changes with amazement, often wishing that he could adjust like his four-legged friend. But now winter was behind them. They had left the plains and were crossing the tallest mountain the boy had ever seen. And below them was the most spectacular valley. The pair had fallen in with a band of Shoshones who had brought horses to help the river travelers cross the great mountains. This place was as beautiful as any the boy had ever seen. It might even be a good place to stay for a while. For reasons inexplicable, both the horse and the boy seemed to feel something special for this land. Little did the young stallion know that he was not at all far from where his most distant ancestors had begun their evolutionary journey on earth more than fifty-five million years before.

Across the Alaskan bridge into Asia, on to the Middle East, then to Spain, North Carolina, and now back home, where it all began.

15

And Nature Dies

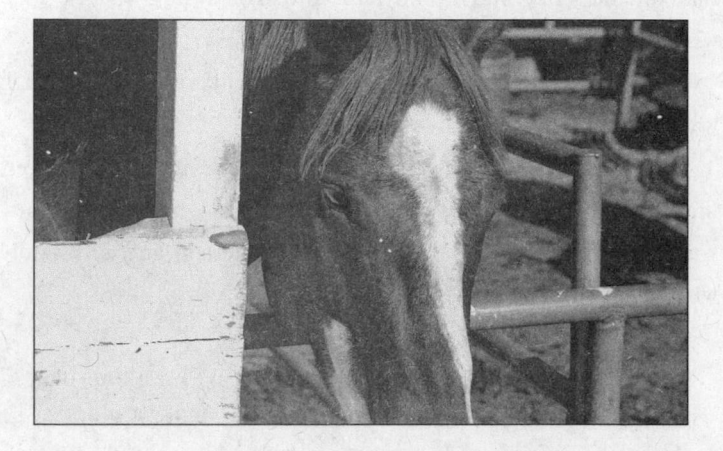

"I'm sorry, I don't care what you say, I'm not watching my horse stand around out there freezing when I can put a blanket on him and he'll be warm."

"It's only fifty degrees," I mumbled.

"I'm cold, so *he* must be cold."

"Are you a horse?" I asked.

She rolled her eyes and walked away. I have always believed that the shortest distance between two points is a straight line. I was about to learn that this isn't always the case.

"What possible harm can come from it?" she tossed over her shoulder.

I followed.

"The blanket disturbs his natural thermoregulatory system." I was spewing research I had learned of only the day before. "He can't grow his winter coat with a blanket on. And his system works on the whole horse, not parts. When he's covered with a blanket, he's half warm and half cold. His system has no idea what to do."

"Well, go talk to his system, not to me. Cold is cold and warm is warm."

"For millions of years the horse has done just fine without blankets," I crowed. "When you disturb his natural systems, you're messing with nature, with his genetics, and ultimately with his health and safety."

She turned on me.

"Do what you want with your horses and leave me alone, okay?"

Clearly my bedside manner needed work. I stood there frustrated, with no clue why this newly discovered information was of no interest to someone who I knew truly cared for her horses. I would never make it in politics.

German philosopher Arthur Schopenhauer said, "All truth passes through three phases. First, it is ridiculed. Second, it is violently opposed. Third, it is accepted as self-evident." Knowing that didn't make me feel any better. Or understand the mystery of this woman's reaction. It was merely preparation for what was to come.

I was standing by a small arena at a local horse club event. The woman next to me was the mother of a teenager trying out a beautiful gaited pinto horse in the pen. The horse was prancing, lifting his legs high, and looking very spiffy. At one point the owner said proudly, "And he's totally barefoot."

"Wow," said the mom. "Just think what he'd look like with shoes on!"

The owner had the grace to ignore the comment. I didn't.

"Why ever would you put shoes on him? He's happy and healthy and looks great."

"Oh, if you compete on him, he *has* to have shoes."

"Why?"

"He just has to."

"Why?"

"Well, it's probably in the rules."

"Barefoot horses compete," I said.

"Well, trust mc, he needs shoes. There are special shoes that make gaited horses prance higher."

"Oh, so that's something *you* want, not something the horse wants."

There I go again, I thought. Like a reformed smoker. My presentation definitely needed work.

"Doesn't hurt him," the woman said.

I took a deep breath, swallowed the words that were threatening to escape, and handed her a card with our website on it, suggesting that she look at some of the new research documented there on how shoes affect the health of a horse's hoof. Then I mustered a friendly smile and left.

I told Kathleen the story.

"We're going to lose every friend we have if you don't shut up," she said.

"I never saw this woman before," I whimpered. "She's not a friend."

"You know what I mean. People don't want to hear this stuff."

"Do you disagree? Is it incorrect information?"

"Of course not," she said. "But . . ."

"But what?"

She just looked at me for a long moment.

"Think about Skeeter," I said. "And how much happier and healthier he is since you brought him here."

"I know," she said. "I know. You're right, but it's so frustrating to have people's claws come out like they do. When you slap people in the face with the notion that they're doing something wrong, the natural reaction is always going to be to defend themselves. You do it yourself."

"I do?"

Her mouth dropped.

"*All* the time."

"When?"

"Last night when I told you that paragraph in the book was not good. You bitched, and screamed, and got ugly . . . and then got up this morning and changed it."

I thought about Schopenhauer's quote.

"Phase Three," I said.

We both smiled.

It kills me to find that so many horse owners make decisions based upon what they *think* is best for the horse without really doing the research. Or use human criteria to make the decision, not equine criteria. And the horses suffer as a result.

It began, I suspect, back when man first decided to dominate one of these thousand-pound animals. Working strictly from fear, with no comprehension of the possibilities available when the horse is given choice and a relationship is built, he must have believed that force was the only consideration for domination. And if one doesn't dominate a beast so large, surely the beast will do the dominating. And hurt you. And ignore your will. This was the lesson taught by the old cowboy from whom we had bought Mariah. And, in days of old, such an attitude was just fine with everyone because that's what man did. Dominate.

Genetics again.

At the expense of the horse.

A well-known clinician was asked to respond to a question from a woman who wanted to enter jumping competitions with her horse. It seemed that whenever she went to such a competition, her horse refused to jump. She wondered if it could be a negative reaction to being around so many horses or being inside a big, noisy facility. "Please tell me what to do," she begged. I never saw the clinician's response, but I hope he told her to begin by evaluating whether or not her horse *liked* to jump. Before being asked to do extreme competition, shouldn't a horse have some inherent desire to do it?

Like the new Benji. She enjoys performing, reaching, figuring things out. We have other dogs who could care less. And one who would be totally intimidated by the workload. To put that dog through a movie production would not only be a disaster, it would mean massive stress for the dog.

When we were searching for the new Benji in shelters all over the country, I looked for a dog that not only resembled the original Benji, with those famous big brown eyes, but was smart and intuitive, and, most important, loved to work, loved to please his master, loved to take on the kind of long and difficult chores that are always present during the production of a movie.

It's true that a horse who doesn't like to jump, or rope, or cut cattle, or run barrels, or race can be made to do it. If the horse is strong and athletic, he can probably be made to do it pretty well.

But doesn't it stand to reason that if the horse really enjoys doing something, he will do it better than if he doesn't? And he'll be a happier horse. And he and his human will have a better relationship. And he'll be without the stress that comes from doing what he hates, or what he is mortally afraid of doing. Which means he'll live longer.

And if participating in the competition is of his choice because he *likes* doing it, and if he's been taught well, there will be no need for force. Or cruelty.

What kind of force or cruelty?

When Dr. Matt was out to vet check Kathleen's new horse, Skeeter, he ran his hand gently across the big palomino's rib cage. There were thirty or forty small dimples in the coat and skin. On each side. Dimples like you might see in someone's chin.

"Know what those are?" he asked.

"No idea," we said.

"Internal scars from spurs."

Our mouths dropped open. And we choked back tears. Skeeter is a beautiful eighteen-year-old quarter horse, who has done some dressage but was mostly used as a roping horse in competitions.

This is the sweetest horse you could ever want to meet. As willing and well mannered as Cash. Without a mean or ornery bone in his body. Yet somewhere back in his history, some human was so obsessed with ego that spurs were used violently enough to leave more than eighty scars in his sides. There is simply no acceptable excuse for that sort of treatment of another living being.

Either Skeeter didn't like what he was doing and had to be forced to do it with extreme spurring or he hadn't been well trained and therefore didn't understand what he was doing well enough to do it without injury.

From the day he arrived at our place, watching his expressions has tickled me to laughter. He loves his new life, but his scrunched eyebrows and big questioning eyes seem to belie a fear that any minute he might be awakened from a spectacular dream.

He won't be. That's our promise to Skeeter.

He joined the herd in our natural pasture, and for a while he seemed to not know what to do with all the space. He would just stand around, bug-eyed, and watch the others. Eventually he assimilated, but he still seems to be amazed that life can be so good. And we'll not take that away from any of them.

"But ugh, being in a natural pasture 24/7, they stay so *dirty*."

"Horses do like to roll."

"I like my horses clean."

That conversation actually happened. What was best for the horse was of no concern. What the human liked was the issue.

There's a photo of a stalled horse in *Horse & Rider* magazine. Below the photo is this headline: I AM CONFINED . . . THEREFORE I AM AT RISK. The subhead: "Confinement-related stress can cause stomach ulcers in your horse—in just 5 days." It's an ad promoting medication for the ulcers. Not a word about eliminating the source of the stress or any discussion about other health problems that could be caused from a stress so tormenting that it produces stomach ulcers.

That same stress, according to Dr. Katherine Houpt, a leading animal behaviorist at Cornell University, in an interview in the same magazine, is responsible for virtually all of the so-called stall vices. Pawing, weaving, head bobbing, stall kicking, cribbing, wind sucking, wood chewing, and tongue lolling are all a direct result of the horse's not being out with the herd, moving around, munching most of the time, with lots of roughage in his diet. Getting the horse out of the stall is all it takes. According to Dr. Houpt, these "vices" have never been observed in horses who live as mother nature intended. Considering the number of products being advertised to "solve" these problems, one has to offer kudos to *Horse & Rider* for having the courage to even publish such an article.

Humans trim their horses' coats in winter to keep them *looking good* for the show ring. This undermines the horse's ability to protect himself from the cold. Wearing blankets does the same thing. As does living in an enclosed barn, especially a climate-controlled barn. Yet this is the lifestyle of the majority of horses in the United States. Such accommodations, usually a small stall, also remove the horse's ability to fulfill his need to move, which affects his feet, circulation, immune system, and general health, as mentioned earlier. And it takes him away from the herd, which causes more stress. And makes him unhappy. And, as you've also read before, often leaves him standing in his own urine and poop, which also adversely affects his feet, circulation, immune system, and general health.

And so it goes.

Humans have the most extraordinary ability in the world to rationalize.

When Dr. Matt came by to do the vet check on Skeeter, he was wielding a huge oval-shaped stone, much larger than a softball. It looked like a rock out of our pasture . . . but it was out of a horse. A horse who had been fed a diet of 100 percent alfalfa. Alfalfa is not grass hay; it is a legume, and alfalfa grown in the southwestern part of the United States is very high in magnesium and calcium, the building blocks of stones. Per pound, there is actually four times the amount of calcium in our alfalfa than the average horse needs. An epidemiology study at the University of California, Davis concluded that 95 percent of horses referred for enterolith (stone) surgery ate a diet made up of more than 50 percent alfalfa. Yet Dr. Matt estimates that a majority of horse owners in our area feed 100 percent alfalfa. He does surgery regularly to remove these stones, and several times a year he has to put horses down because they're too far gone—the stones too big, like the one he was holding in his hand.

Why do they do it? With all the information available about equine nutrition, why would anyone in this part of the country feed straight alfalfa?

"We work our horses hard. We're riding three or four times a week. We need them to perform. Go fast. Alfalfa keeps them *hot*."

Is there another way to keep a horse's energy level up without ingesting so much calcium and magnesium? Of course there is. But it might cost a bit more. Or take a bit more effort.

Eighteen-year-old Skeeter was on straight alfalfa when he came to us. We tapered him down to 20 percent alfalfa and 80 percent coastal Bermuda. He lost a bit of weight, apparently accustomed to the energy and protein in alfalfa. So we studied a bit, spoke with Dr. Matt, and added some rice bran to his diet. A bit more trouble, maybe a tad more expensive. But his weight is fine, he's a happy camper, and he's not getting those excessively high doses of calcium and magnesium that can cause huge life-threatening stones.

Living in the wild, horses eat a varied diet that usually includes legumes like alfalfa, but the majority of the diet will be of the grass family. Dr. Strasser reports that horses have an instinct for what medicinal properties certain plants have, and they know when there is a need for such natural medicine. In the wild, for example, they'll go for thistle when their liver is irritated. *Before* it becomes a problem.

In conventional boarding, horses are usually fed two or three large meals a day, leaving lengthy periods with no ingestion. A horse's stomach is very small in relation to his size, perfectly suited for virtual continuous ingestion of small amounts of vegetation; and he needs to be continually processing roughage for his digestive organs to be in good health. When there are hours between feedings, which often consist of grain or pellets, without roughage, his body cannot function as it was designed to function. It makes him crazy, contributing to the stress of being confined, and helping to create those so-called stall vices.

The horse has taken care of himself for millions of years. It only requires a bit of research, effort, and imagination for us to figure out how to help him replicate as much of that life in the wild as possible that will cause him to be happier, healthier, and live longer.

Why would we not?

16

Love Is the Gift of Oneself

The young Powhatan sat on a boulder watching the golden stallion play with one of his foals in the herd. The boy had grown tall, and lean, and muscular, and was no longer a boy. His ability to run, and hunt, and think had endeared him to the Shoshone chief and he had become a special adviser to the great leader of the tribe.

It was his twelfth time to return to this spot where he had given his blessing and prayers to the stallion and encouraged him to follow his calling and join the herd. It had been a sad day for both of them, but the young man had well understood the great stallion's wish for fulfillment, for he, too, was a man now.

When they had first encountered the herd, the young man could sense the stallion's emotions and knew what he wanted. He talked to him, his head buried in the stallion's mane, and prayed to the great spirit to take care of him. Then, finally, he stood back and told him he was free to go. The horse stood for a long moment looking deeply into the eyes of the human with whom he had shared so much. Then he wheeled and raced away down the hill.

The reigning stallion of the herd was actually glad to see him, for he was old. There was no fight for supremacy, no confrontation, just a passing of the torch, and the old gentleman was allowed to remain with the herd.

Since that day, the Powhatan always returned on the first day of snow and when the spring flowers began to bloom. And the herd was always there.

The matriarch began to move her charges into a stand of trees, but the stallion remained. He turned toward the hill and his friend sitting on the boulder. Then he came, just as he had done each of the times before, trotting up the rise, only stopping when he was nose to nose with the young man. They nuzzled for a bit, then talked about things. Although neither probably fully understood what the other had said, the moment, the history, and the bond were well understood. The Powhatan was glad that his adopted tribe wasn't around to see the tear drop from his cheek, because he had no desire to hold it back. He loved this horse more than he had ever loved anything. Which is why there had been no choice but to release him.

17

Horses Aren't Us

For three hours I sat on a big, hard rock watching four of our horses investigate their new world. Our natural pasture was, at this time, unfinished, but we thought it wise to put it to the test before locking down the final touches. The horses paraded from the north end to the south, frolicked and played and moved each other around. They ran up and down the hill, nibbled on stray weeds, and otherwise had a fine time being out in the wild, so to speak. Everything went very well.

Until we left the pasture.

Kathleen and I decided to allow them to make the short walk back to their smaller turnouts untethered. These horses mind really well and the property is fenced, so they couldn't wander off. We felt it would be a happy extension of their new freedom.

You guys just come on along. Follow us.

Three of them did.

I was walking out front and didn't notice that Cash had paused to nibble a weed. When I discovered that he wasn't with us, I walked back and called out to him. He looked up, realized he was all alone, and, as Pat Parelli says, went totally right brained. He leaped from a dead standstill to a full gallop, racing right past Kathleen and me, to catch up with the other horses. None of that would've been a problem were it not for a small section of concrete driveway that must be traversed on the way home.

A small section that makes a tight U-turn.

Going downhill.

When Cash hit the concrete at full stride, trying to make the turn, his feet flew out from under him and he landed with a loud *kerwhomp* on his left side, leaving hair and bloody skid marks as he slid.

I was frozen in place, horrified.

Cash is the most athletic horse we have. The smartest. The most well mannered. The most loving. And the most accident prone.

He's been kicked in the head and bitten, has sprained a foreleg ligament, and today, as I write this, he has a sprained right rear fetlock in a soft cast. It's not unlike a twisted ankle, probably from stepping on a loose rock or in a gopher hole in the new natural pasture. Shortly after he arrived at our place, I watched him go straight up a rock face, petrified that he was going to wind up skiing all the way back down, maybe on his back. He managed through that one without even a scratch. The fall on the concrete·

left him scraped and bloody in several places, but nothing was broken or seriously injured. We were lucky.

There's a piece of me that wants to lock him up in a padded cell and never let him venture outside. I know that if anything serious ever happens to him, in some way it'll be my fault. I wasn't a good leader when I let him get to that rock face, or when I allowed him to come back from the pasture on his own without a lead rope, or when I let him be in the same stall with the horse who kicked him. And, apparently, just letting him be in that steep, rocky, natural pasture leaves him open to injury.

But would I take that away from him?

Should I take that away from him?

There lies the toughest part of entering the world of what is truly best for the horse.

Us.

We aren't horses; and if that's true, it stands to reason that horses aren't us. So we mustn't treat them as if they were. We must be able to rise to the occasion and accept the facts:

What is freezing to us isn't to them.

What is safe and comfortable to us isn't to them.

What is warm and cozy to us isn't to them.

What breeds trust and respect for us is different for them.

There are folks who believe that the best care for the horse is the safest care, no matter how miserable the horse's days might be, no matter how many years are being cut off his life because of a sedentary, stress-filled lifestyle—in short, no matter what.

Cash hates—I mean *hates*—being cooped up. I think the second-happiest day in his life was when he was turned loose in that natural pasture. The happiest was the day his shoes came off. Until that day, I had never actually seen a horse smile.

What's best for the horse is almost *never* what we humans think is best for the horse. As much as we'd like to believe that horses are like us, they aren't. We must always ask ourselves, Is this what *I* want or what my horse wants? Is this truly best for my horse?

To ask those questions might involve embracing change. The awful *C* word. But it's a lot easier to accomplish what's best for the horse than most people think.

Our natural pastures, for example, were relatively simple and inexpensive to put together. We have three: two small ones and one larger, which is about an acre and a half. They're all dirt, rocks, and steep hills.

Oh my, shouldn't a horse pasture be flat, without any dangerous obstacles?

You mean boring?

Well, on steep rocky hills a horse might get hurt.

Tell me about it!

So it is dangerous.

The horses prefer to call it interesting.

Oh my.

Actually, the steep hills and rocks are good for the horses' bare feet and teach them to pay attention to where they're stepping, which translates well to the trail. The level of exercise is better, thus circulation and muscle tone are better, and stamina is improved.

The large pasture, where all six of our horses now spend most of their time, is enclosed with an inexpensive electric fence. In the morning each horse receives a bit of Purina Strategy, about half a scoop as a supplement, a small bit of alfalfa, and approximately six pounds of Bermuda grass hay, a total of approximately thirty-six pounds for the herd scattered in fifty to sixty small piles all around the pasture in a big circle. That keeps them on the move most of the day, checking first this pile, then that one. Night feeding is the same. The smaller pastures are odd shaped, approximately fifty by sixty feet, and would work very well even if that was all the space we had.

Our horses are out 24/7 with a single exception. If the temperature drops below 42 degrees *and it's also raining* we'll bring them into those cute little open stalls mentioned in chapter 2, which are half covered, with one closed side facing the usual weather. This allows the herd to stay dry if they so choose. According to most advice on this subject, neither rain nor freezing temperatures matter much to a horse unless they both occur at the same time. When the coat is soaking wet, it might not be able to protect against extreme cold, so we take Dr. Matt's advice of better safe than sorry, and give the horses a choice of whether to stay under cover or not. We have a couple who never seek shelter. Go figure.

The pasture is very close to these horses' natural habitat. It certainly costs less than building a big barn and can often be set up in about the same amount of space, depending upon the number of horses who will occupy. For boarders, more and more natural pastures are popping up all over the United States. There's even a racing and hunt club in the United Kingdom that requires all club members' horses to be barefoot and live in the pasture with the herd. The name of the club is Horses First, and its horses are winning races all over the country.

As mentioned earlier, I sometimes want to put Cash in a padded room to keep him from hurting himself. Wrap him up in a thick, fuzzy blanket. Feed him warm tea and hot chocolate. But watching him pace in his stall awakens me from this human delusion. Until his fetlock ligament heals, he'll have to remain confined. But I haven't forgotten how happy he was when he was first turned loose in the big natural pasture. The smile on his face. The prancing. The racing. The kicking. He was eleven hundred pounds of pure frolic.

I have no choice but to let him go back as soon as he's well.

Because I love him.

18

The Bond

The young Powhatan knew his horse.

Inside and out.

They had spent years together as each other's only friend. They spoke the same language. They understood each other's touch. The Powhatan knew, intuitively, what the stallion needed when it came time to release him. And the great stallion knew that it was okay to leave. It was the right thing. The strength of the bond between these two is rarely even dreamed of in the domestic horse world. And that's very much a shame because it's truly amazing how little time it has taken Kathleen and me to nurture such a bond with our horses. And how consequential that bond has been to everything else that we have done.

There are those who believe it's not possible to bond with a horse. Dogs, yes. Cats, maybe. But not horses. Yet I experience this bond with every member of our herd on a regular basis. And sometimes the things they do to confirm it can make my day, or my entire week.

Dr. Marty Becker says in his wonderful book *The Healing Power of Pets*, "We should recognize the bond for what it is—living proof of the powerful connectedness between mankind and the rest of the animal kingdom. And the element of this powerful relationship that has always impressed me the most was the importance of nurturing another creature."

I wonder if those who don't believe that you can bond with a horse also dismiss Dr. Becker's statement about the importance of nurturing another creature.

Just last week I was hauling my tripod and video camera down our very steep pasture, struggling a bit because of sore ribs, a remnant from my fall off a ladder. I was moaning to myself about it as I set up the camera to videotape herd movement. Five of our six horses would soon begin their meandering climb back up the steep grade toward where the camera was set up. Only Mariah had remained at the top.

After a moment, I heard her shuffling up behind me. She paused at my back, lip-nibbled my shirtsleeve, and then the most amazing thing happened. She nudged her nose between my arm and my ribs and pressed her warm muzzle softly against my rib cage. Precisely where it was hurting. She didn't move for minutes, until I had to shift position to start taping. It was a moment I didn't want to end.

How did she know?

Moreover, why did she care?

This is the horse whose relationship with humans was a blank stare when Kathleen and I first met her.

This is the bond.

Kathleen and I spend regular time in the pasture, without agenda, to foster this bond. And to learn about our horses. The relationship, generated with Join-Up, which gave the horses the choice of whether or not to be with us, continues to mature because of our time in the pasture. And we become better communicators. Everything about the relationship gets better.

Time in the saddle, in the arena, and on the trail are important. But I believe the most important time is in the pasture. Just hanging out. It has done wonders for us and our horses.

It continually strengthens the bond and our relationship.

It teaches us about the horse, his habits, his language, his individual personality, and his genetics. How to read and understand what makes him tick.

It strengthens our leadership and the horse's respect.

It dispels fear, both ours and theirs.

And it breeds confidence.

None of that can be injected, like a flu shot. It doesn't come as a flash when we wake up one morning, no matter how much we wish that it would. And even though books and DVDs have certainly crammed us full of insight and knowledge, they cannot replace the benefits of experience that come with being there, doing it, absorbing, learning firsthand.

Simply hanging out in the pasture, observing, studying, interacting at the horse's discretion, has taught us so much. That's why we wouldn't pay someone else to regularly feed and muck, even if we could afford it. Yes, there are mornings when we'd love to sleep in, especially in the summer when the kids are out of school and we don't have to wake up before dawn anyway. But doing our own feeding and mucking guarantees no less than a couple of hours a day with our horses. Over time, those hours help to dissect and internalize each horse's individual personality, which determines how leadership is expressed in different ways to different horses. It provides insight into how weather affects their behavior. It has taught us, virtually by osmosis, how subtle our language can be, or not, with each unique horse. And it continually confirms us as members of the herd.

"I just never have enough time," one woman said to me.

"Then maybe you should acquire something that doesn't depend upon your leadership, relationship, compassion, and understanding for its health and happiness."

I didn't really say that, but I thought it. You see, I *am* learning.

Spending time with the horses also reminds us to always be thinking ahead, questioning, anticipating what could happen or go wrong by doing things this way or that.

Only yesterday, with enough mileage in the pasture to know better, I was accidentally knocked down by big, muscular Pocket. She's our *other* paint. Besides Scribbles. Major big. Not so tall, just *big*. Probably pushing twelve hundred pounds.

· The incident wasn't her fault; it was mine. And the time I had spent in the pasture told me so immediately, even as I sat on the ground staring up at her. Still, my first reaction was anger. I wanted to yell at her. And I know folks who would have. I know people who would whip any horse that would do what Pocket had just done, with no thought to understanding why it had happened. They would rather have a horse who is totally afraid of them than enjoy a bond and relationship, and be truly responsible for their horse's leadership.

This is how it unfolded.

Each of our horses has a small feed tub in the pasture and they all know which one is theirs. With the exception of Skeeter, which is another story, it takes each horse approximately the same amount of time to eat the first course, the appetizer—a half-scoop ration of pellets. Next on the menu, the antipasto, is a small amount of alfalfa hay, less than half a flake per horse, scattered in ten to eleven small piles, all in relatively close proximity, at the top of the pasture hill. The manner in which we spread it ensures that no one horse can dominate more than his share of alfalfa and no one horse gets eliminated from the game of musical chairs that follows.

Yesterday, for reasons I don't even remember, I put several of the alfalfa piles much closer together than I usually do. Four horses bunched up on the same piece of rock, all vying for as much of the booty as each one could get. I should've moved out right then, but I didn't. I continued to pull apart the flake. When the dominant Scribbles took a nip at Pocket, she leaped out of his way, bumping Handsome, who whipped around, threatening a kick. She had nowhere to go but straight toward me. She was otherwise surrounded by hostile troops. She tried to miss me, and actually just brushed my shoulder, but with force enough on unlevel ground to sit me down. There was simply too much congestion for safety and decorum. Especially when a nip and a threat had spiked her adrenaline. I should've known better. Now it's well implanted in my brain by the bruise on my butt.

One look at her face, however, confirmed beyond doubt how she felt about it, and perhaps told of a bit of history.

Omigod, what have I done?!

She is usually the first horse in the herd to come greet me when I enter the pasture. Our bond is strong. But I couldn't even get close to her for several minutes, as if she were expecting punishment. Or was really, really embarrassed about it all.

When I did finally get close, I rubbed her forehead and told her everything was okay. Well, except my butt.

And I promised to never again place alfalfa piles that close together.

One cold blustery day, Scribbles, the paint who is dominant in the herd, was acting out a bit more than he normally does in asserting his God-given right to eat first, to be the head of the table, so to speak. I was entering the pasture with pellets, heading for his tub. He was flipping his head at this horse and that, and quite without thinking he turned and flipped his head at me and launched a tiny kick. He was ten or fifteen feet away, not within striking distance, nor did he have any intention of striking. It was just a misplaced dominant gesture in the middle of his dance. He got carried away. I stopped walking, swelled my body like a balloon, looked him straight in the eye with eyebrows raised and one finger pointed straight at his forehead . . . and just stood there.

That doesn't work for me!

You could literally see the *gleep* on his face. I could almost hear it. It was all I could do to keep from laughing out loud.

I stood there for at least thirty seconds, way longer than Scribbles could normally stand still when food was on the way. But stand still he did. And his head dropped almost to the ground, a very submissive posture.

Boy, am I sorry. Didn't mean to. Really! It's that cold wind. The devil made me do it.

I walked over and rubbed his forehead, and then proceeded to his feed tub.

Without a word spoken. Without any physical threat.

The relationship, the bond, carried the load.

It happens with every horse in the pasture. Not every day. With some, not even every week. But it's there, and it makes everything else so much better.

In addition to the time we put in feeding and kicking poop, Kathleen and I usually spend no less than a couple of hours a week just sitting on a rock, watching the horses interact. Or having a cappuccino in the morning or a glass of wine in the evening on our deck just off the pasture. It's fascinating to see what the flick of an ear, or a nose, can mean. To watch relative body positions, and how they relate to causing movement. To notice when the dominant horse will let something slide, and when not. And why. Digesting all this has seriously improved our equine vocabulary, and our natural responses to the horses. How, for example, a slight change in body position can vary the message being sent. How a simple flick of a finger, when it's accurately placed with the right attitude, can accomplish what once took a broad arm motion.

And because they are horses, our every improvement in understanding their nature creates a greater respect, and a stronger bond, which, by the way, is not just an equine characteristic.

The experience became priceless once we realized that, most of the time, when a horse is refusing to do what we're asking, it's only because he doesn't understand what we want. The better we communicate in his language, the more he will do willingly for us—and the stronger the relationship.

This has been our way of life since we finished the natural

pasture almost a year ago. Sometimes we forget how other horses live. I recently visited a traditional boarding stable, a fancy one, heavily laced with dressage and show horses. I was struck with a pang of sorrow. I wanted to race through the place and pop every latch on every stall door. And pull every shoe. And rub every horse. And Join-Up with them. And listen to them.

"How often does this horse see her owner?" I asked. The horse was a beautiful thoroughbred cross, with very sad eyes, sort of glazed over. She was pacing, back and forth, back and forth. I stood at the stall door for a moment, but the horse never looked at me. Just paced.

"Oh, at least once every weekend. Sometimes twice."

"Who feeds her?"

"We do. Our staff."

"The stalls?"

"Our staff. Twice a day."

"Does she get turned out?"

"Oh, absolutely. Four hours every day. Guaranteed."

"Guaranteed, huh?"

"Absolutely."

"Turned out with other horses?"

"Oh no. She might get hurt."

"Does she have a trainer?"

"Uh-huh. Comes on Tuesdays and Thursdays."

"How does the horse know who her leader is?"

"Oh, the trainer's the leader, if you want to call it that."

"And the owner?"

She looked at me like I was an idiot. I should know the answer.

"The owner is the rider in the shows," she said. "And, of course, she writes the checks."

"I see," I said, and walked on down the row of stalls.

In one way or another they were all the same. Horses penned up, away from any semblance of a herd. Showing stress in one way or another. Pacing, weaving, chewing, cribbing. One was kicking the wall. The stalls were all filled with a bedding I didn't recognize, presumably to absorb the pee, but the odor was still present, digging into their lungs. And most of them wore metal shoes.

"I notice that this one is barefoot," I said. "He doesn't have shoes."

"Strange owner," she said. "He doesn't understand that a horse has to have shoes. That's the way it is."

"Do you own horses?" I asked.

"Don't need to. Just look around."

It all reminded me of my sailing days, back when Benji was at his peak. I had a sailboat in a Fort Lauderdale marina, and I was always amazed that so many people had huge boats—mine wasn't—yet those huge boats never left the slip. The owner would pay the boarding fees, pay people to keep the boat clean, to run and service the engine, and would come down once a month and use this big, expensive yacht as a hotel room. Never go out. Never enjoy the boat as a boat. It was just a place to hang out and sleep. And be proud of.

As frustrating as that seemed to me at the time, I suppose the good news was that it was a *boat*. An inanimate piece of fiberglass and machinery. And if the owner wanted to pour his money into that . . . well, it's his money.

But a horse is not a boat.

A horse lives and breathes, and has feelings, and worries, and a genetic system that has evolved over millions and millions of years. A boat doesn't pace and stress when it's tied in a slip. It doesn't have a herd that it should be with. It doesn't worry about being attacked by everything that blinks. A boat doesn't need to be out where it can move around all day and night. A boat doesn't feel happy or unhappy, or healthy or unhealthy because it is or isn't being ignored by its owner.

A horse does.

Mariah did, once upon a time.

But she doesn't feel that way anymore.

Ask her and she'll tell you which way is better.

19

Feelings

The stallion stood in the flat, searching the rise around the big rock. His friend had not shown up and it was now the day of the third snow. Winter was on its way and he sensed that it was going to be severe. Never had the snow come so early, nor the second and third snow so quickly. The matriarch was restless, wanting to move their charges south before the weather made it impossible.

The stallion left the herd and trotted up the rise for a closer look, searching the trees along the ridge and scanning every inch of ground down the rise. His friend had never missed the first snow. He had no way of knowing that the Powhatan warrior had been badly wounded in a battle between his adopted tribe and the neighboring Blackfeet. He had fought valiantly to help the Shoshone protect their hunting rights so they could feed their families.

The snow was coming harder now, blanketing the rise in white silence. The huge golden horse looked to his herd and back up the hill. He paced nervously, and the snow fell faster. He was just retreating to the herd when a movement caught his eye, high up the rise. A flutter, a shadow in the snow. He strained to see through the filter of white. But it was gone. Then back. Another movement. He was sure of it. He trotted up the hill, gaining speed as he recognized the shape on the ground as human, his human. He slid to a stop and looked down on his friend, crumpled in the snow, a trail of red stretching out behind him. The Powhatan was trying to pull himself toward the rock, the meeting place, but his strength was gone and he was shivering uncontrollably. He saw the great stallion's feet and strained to look upward. The sight of his friend vanquished his pain. A smile spread across his face and a tear slipped down his cheek.

The great horse dropped to his knees and stretched out on the ground next to him, trying to warm him. His friend wrapped his arms around the big stallion's neck, clinging to him as if he were life itself.

And then he was gone.

The stallion's scream shattered the surrounding silence, again and again. But he didn't move. The matriarch appeared, and others from the herd, but they stood quietly and watched the falling snow cover this human warrior who meant so much to their protector. Finally, the stallion climbed to his feet and gazed down at the man, the human, with whom he had come so far. He reached to the ground and nuzzled away the mound of snow hiding the man's cheek. The warm nostrils of this elegant stallion breathed in one last remembrance, then he lifted his head to the trees and bellowed as if his heart were being ripped apart. Then he turned and raced down the rise toward the south, flying like the wind. For hours he ran, until he could run no more, and he lay down and went to sleep.

When he awoke, he was surrounded by his herd, who had followed him south and almost kept up with him. The stallion would never again see the valley of his friend. The grasslands would turn into hard earth and rock, and then prairie, but his herd was fit and their feet were like rocks themselves, and the journey would be made without incident or injury. When they reached a fine valley with good grasses, they stopped, and there the stallion's progeny would live for ten generations.

It would be more than a hundred years before any of them would ever meet another human. But, somehow, the memory of how much humans can mean to the horse would not be lost or forgotten.

20

Sonny Boy and Painto

We know it's there. That relationship between horse and human. All horses. Any horse. If you know his language, understand what makes him tick, what makes him feel safe, that horse will allow you into that special place called relationship. Some feel the connection can be spiritual, even mystical. Those who believe the horse is inherently mean and must be dominated have sadly missed a very important fork in life's road. An editorial in the *New York Times* following the euthanasia of Barbaro put it this way: "You would have to look a long, long time to find a dishonest or cruel horse. And the odds are that if you did find one, it was made cruel or dishonest by the company it kept with humans. It is no exaggeration to say that nearly every horse is pure of heart."

In all his years, I believe Monty Roberts has only encountered one horse with whom he wasn't ultimately able to establish a relationship.

"Are you saying you should try to establish a relationship with every horse you encounter?" I was asked not too long ago.

"If you intend to spend any time with him, yes," I answered.

"Even borrowed horses, or rental horses?"

"If you care about the horse as another living being, yes."

"Is it really worth it? Does it make a difference?"

"To you, or the horse?" I asked.

He shrugged, not sure how to answer.

"Either, I guess," he said. "Both?"

I told him this story.

Kathleen and I were headed for a three-day trail ride west of Albany, Texas, riding borrowed horses because it was too far away to bring ours from home. This would be the first time since our nosedive into the horse world that either of us would be on horses other than our own. Horses we did not know. Horses we had not schooled. Horses with whom we had no relationship, who had never heard of Join-Up. And there was no facility to accommodate such a silly notion.

"I'm told these are good horses," I said to Kathleen. "Well trained. Calm. Let's just go and have a good time. Put the whole relationship thing aside for three days."

"I can do that," she responded, "but I doubt that you can. It's your scorpion."

There's an old story about a scorpion begging a frog to swim him across the river. The frog calls the scorpion nuts. *We'll get in the middle of the river and you'll sting me!*

Why would I do that? the scorpion replies. *I'd drown with you.*

That made sense to the frog so he agreed to swim the scorpion across, and sure enough, right in the middle of the river, the scorpion stings the frog.

Now we're both going to die! screamed the frog.

I know, said the scorpion. *I just couldn't help myself.*

"I can help myself," I promised Kathleen. "I can do it."

One of the reasons I really wanted to go on this trail ride is that we had been wrapped up in the world of horses for more than a year and Kathleen still harbored a bit of fear, rising on occasion, just under the surface. Especially when riding faster than a walk. Her prior experience with horses, before we owned our own, was, in her own words, "from the ground looking up." Four trail rides as a teenager on a horse named Jack. She was dumped all four times. Unbeknownst to her at the time, Jack had also dumped a jet-fighter pilot. Those were her last times on a horse until she gave me the trail ride for my birthday. I was convinced that mileage was all she needed. Three days of riding, all day, every day, would take her light-years down the road toward confidence. I was certain of it.

She was now much more experienced, but she was also putting up a good front. Every time she climbed into a saddle, her insides were churning, betrayed by the number of times she quoted clinician Linda Parelli saying, "Whenever you feel the least bit uncomfortable, just get off. Right then. Immediately."

Some say stay on a bit longer. I discovered years ago that you cannot rid yourself of any fear if you don't push through it. If you don't stretch yourself. The trick, then, is how to do that while staying out of harm's way. Keeping yourself safe. Only then will you be calm and focused enough to actually learn and become confident in your skills.

But how do you do that if you have a crazy, high-spirited horse?

The simple answer is: Don't have a crazy, high-spirited horse if you're a beginner.

I've watched dozens of clinician DVDs and a few live clinics, and I'm continually amazed at the horses many of these scared-out-of-their-minds beginners (and not-so-beginners) show up with. You only lose your fear through mileage—time and experience—and you're not going to get much mileage if you're afraid your horse is going to kill you. Making matters worse, when you're afraid, your adrenaline is up and your horse can read that, and his adrenaline will rise as well. Monty Roberts says, "Adrenaline up, learning down. Adrenaline down, learning up."

Pat Parelli tells about one of his clinics in Australia where he asked everyone to demonstrate their horses so he could get some sense of what he would be dealing with. After seeing each student ride, he told the group that he wouldn't think of riding *any* of the horses he had just seen.

Pat would school them first. Create relationships and safety. And even for a beginner, that can mean progress. But some horses, even with relationship, are too much for some beginners to cope with. Too spirited. Too excitable. It depends so much on the horse, and the human, and the human's fear threshold. Pat Parelli has decades of experience reading horses and communicating with them. For the beginner or early intermediate, there's a much simpler answer to pushing through fears and gaining confidence. Don't let your eye get caught by that young, feisty, wild and crazy horse just because he looks so cool.

It's true.

Kathleen is proof.

She learned nothing when she was garbled with fear about falling off. She learned very quickly after finding Skeeter.

"But I can only afford one horse," comes the usual response.

"Perfect," I say. "Get a Skeeter."

"No. I want a horse that will be good for me when I become an expert."

"If you start with that horse, you might never become an expert. And if you do, it'll surely take much, much longer."

Everybody, of course, doesn't need a Skeeter. People are different, and have different thresholds. And horses are different. The point is to find the right relationship for what you're doing at the time. My fear threshold was higher than Kathleen's. It only kicked in when it came time for me to canter, to go fast. I put off doing that forever. Cash loves to go fast. I didn't, in the beginning. My adrenaline would soar at the very thought of it. Then, of course, so would Cash's.

So finally what I did was drop down to Mariah.

Literally.

Sweet, tiny, sensitive—did I say small?—*obedient* Mariah. If I was to fall off, the ground was only half the distance of a launch from Cash, or so it seemed. Her speed was much more manageable than Cash's. And because her legs were shorter, her gait was shorter. Never mind that none of that should matter; it did. Emotionally. Pushing through my fear would be a lot safer on Mariah.

Perception.

When it comes to fear, perception is the key ingredient. Just ask a horse.

I knew that the only way I could ever effectively work on Cash's speed was to be able to focus on the speed, not my fears. Develop confidence in the saddle at a canter. "Get my seat," as the clinicians say. My balance point.

Lose the fear.

And lo and behold, when I went back to Cash and asked for that first canter, because my adrenaline wasn't up, neither was his, and he loped away at a very calm, leisurely pace, thank you very much, and stopped when I asked him to. It continues to amaze me how that works.

If you can only have one horse, then make it one horse *at a time*. The *right* horse for *each* time. Then you can push right through your fears and concentrate on learning, and teaching, and becoming confident. Both you and your horse. You can actually build a relationship, instead of just going through the motions out of fear.

Don't fall into that instant gratification bucket. *Here, Mr. Trainer, make Flicka the perfect horse that will love and respect me and do everything I say. Immediately.*

Pat Parelli says that's like sending your spouse to a trainer.

Kathleen jokes that we have six horses: mine, and the five she went through trying to dispel her fears. For various reasons, each horse came up too much for her, or so she thought. Until Skeeter. He had pretty much seen it all in his eighteen years; a seasoned veteran who had been around the block, not much bothered him. His roping career had taken him all over the place, in and out of trailers, arenas, and stalls, all furnished with noise, strange goings-on, and raised adrenalines. He was a good home for Kathleen and, after Join-Up, gave her plenty of calm and quiet to practice her training and riding skills. To build her confidence. He taught her well, and now she rides and works with everyone in our herd, except Cash. And I don't think she's far from taking a turn with him.

If she had continued to work with a young, feisty, easily excitable horse—even with Join-Up and extensive work on the ground building relationship, leadership, and control—I don't believe she would've ever effectively been that horse's leader, be what he needed to respect her. There was too much fear and worry blocking her concentration. It was the wrong horse for her to be riding at that stage of her development.

But with Skeeter, she was able to focus, and teach, and learn. And her fears were all but vanquished when we headed off to Texas for the three-day trail ride.

The ride would put us in the saddle for six to seven hours a day for three days, maybe as many as twenty hours, which probably equaled the total number of hours Kathleen had spent in the saddle during the entire past year. Her rides at home usually averaged fifteen to twenty minutes each.

She needed mileage.

The ride was a benefit to restore Fort Griffin, a Civil War village and compound, put on by our dear friend country-western singer-songwriter Red Steagall. I had dreamed of nights around the campfire listening to Red sing and recite cowboy poetry. My Boot Barn boots were now well broken in and ready for the real thing.

We flew into Dallas–Fort Worth then drove out to the campsite with Red and his wife. The horse I would ride, Sonny Boy, was in the trailer, but Painto, Kathleen's mount, was coming with another rider and would be staying at a different, remote campsite. This was a shame. But, as it turned out, it provided a fascinating study.

Red and I put up the temporary corral, and as we walked the horses down to the watering bucket, Sonny Boy and I stopped, backed up, went forward again, and made several U-turns. I asked him to move his hindquarters, and he swung them left and right, catching on to my cues quickly. He seemed to think it was a game, and enjoyed getting a rub for a job well done. I backed him away with the lead line, rubbed him again, and otherwise began to pry myself into the thinking side of Sonny Boy's gray matter.

"Scor...pi...on," Kathleen chortled, walking along beside me.

I didn't even attempt an excuse. She was right. She knew it, and I knew it. I had to connect, or at least try.

I had Sonny Boy trot circles on the lead line, and change direction, and I ultimately stopped and turned away to see if he would walk up to me. A poor boy's Join-Up. He did. I spent some time just hanging out with him, with no agenda. Just being there. After a bit, he was following me around the pen.

And Kathleen had yet to even meet Painto.

Had Sonny Boy and I actually joined up? Probably not. He had been on a line, not free to roam. Was it better than just walking up and climbing on? Absolutely. We were getting to know each other, and he was learning that I was an interesting, reasonably capable human, not just someone who crawled on for a ride. He was learning that I would give him choices, and could engage his brain, and cause him to move. In short, I was establishing myself in the role of a benevolent leader. Someone Sonny Boy would listen to. Someone he could trust. And someone he would not fear.

We had never met. Didn't know each other. But we both knew that one of the criteria of leadership is determined by who moves who around. Who can walk up with a simple pinning of the ears and move the other horse away from a pile of hay or a watering hole. Sometimes it might require a nip on the butt, or just the threat of a nip. The criterion is the same with the human-horse relationship. I've never bitten a horse on the butt, but when I can—with a mere shake of the lead line, a pointing of a finger, or a touch to the hindquarters—move the horse from here to there, and control his various body parts, it generates respect that translates into trust. I become the leader. This is one of those equine concepts that is difficult for humans to grasp because we attach emotion to the action, when, to the horse, there is none. It's just the way they are, and is as much a part of their nature as eating.

There's no substitute for allowing the horse to make a choice, to choose you to be part of the herd, or choose to do what you're asking rather than being forced to do it. There's responsibility attached to these decisions. Join-Up is the ultimate choice, but not the only one. And good leadership only *begins* with Join-Up. The leadership must continue. To be able to engage the horse's curiosity, his thinking side, while knowing when to back off when he does something correctly speaks volumes to him.

A request, instead of a demand, allows him to make a choice. And when he makes choices to do what you request, you become his trusted and respected leader.

Unlike the human fraternity, every equine herd has a detailed pecking order, from top to bottom. If there are twenty members of the herd, there are twenty places on the ladder, and everybody knows their place. If the leadership is not good, that place is often in question. One horse will move up a notch. This behavior has nothing to do with whether the horse moving up likes the other horse or not. It's just the way it is.

Their way of life.

It's difficult for humans to realize that a horse attempting to wrestle away the leadership role isn't forsaking the relationship. But he's not. He's just rearranging it. Responding to genetic programming. We don't have to understand the emotional dilemma humans would like to attach to such a conundrum. But to deal appropriately, compassionately, with horses we must know that it exists. And we must understand how to use this unique portion of the equine lifestyle and language to enhance our relationship and leadership role. We must *be* a horse.

To every horse, so completely concerned with safety and security, his leader is everything. With a good leader, he feels safe. With an ineffective leader, his genetics leave him but one choice. To *become* the leader. This is true when he's dealing with another horse, or with a human in a herd of two. The minute the quality of leadership takes a turn for the worse, the horse is going to attempt to step into that role.

Always testing.

Checking on our leadership qualities.

Are some horses easier to lead than others?

You betcha. Just like people.

If, however, we have begun properly with the horse, if we're at the top of the ladder, with the trust and respect that goes with the position, the horse will make little more than token efforts to test us, just to confirm that he still has a good leader. But if we allow the horse to take over, he will. From his point of view, he has no choice.

At home, Kathleen and I have spent an enormous amount of time working with our horses on the ground, teaching them to move one body part or another, ultimately with little more than a look or a nudge. This not only enhances our position as their leader, it enhances their desire to please and makes them safer, less apt to be challenging our leadership.

That's what I was working on with Sonny Boy.

And what Kathleen was unable to work on with Painto. When we arrived at the trailhead the next morning, she was presented with a saddled, bridled horse, meeting him for the first time. A sniff of the hand and a couple of rubs on the face were about all she had time for before the troops were headed off down the trail.

Even with so little introduction, her first two days were terrific. Painto was very well trained, calm and responsive to the lightest touch. Kathleen was very happy with him and her confidence was soaring. Not just at the walk. There were many long trots.

"What's a long trot?" I had to ask Red. "Is that referring to the length of stride, or the length of the ride?"

"Yes," he grinned through his sagebrush beard. "Both."

What neither Kathleen nor I was told was that both Sonny Boy and Painto had really strong ties to their pasture mates, both of whom were on the ride. One was being ridden by Red, and the other by Jimbo, Painto's owner. But because we all spent most of the time in close proximity, chatting and enjoying one another's company, there were no clues as to what might happen if Jimbo or Red were to vanish into the woods.

Until the third day.

Kathleen's morning began a little differently from the previous two. Jimbo was convinced that Kathleen was in control and doing well, so he had taken off in the lead on Painto's pasture mate while Kathleen was still saddling Painto. He became a different horse in a blink. Nervous. Concerned about his safety. Reactive rather than thinking. His only leader had disappeared up the trail, and it wasn't Kathleen.

She fought him for two hours, afraid to let him go fast enough to catch up for fear she would not be able to stop him. Visions of her teenage experiences with Jack danced in her head. Then the trail boss rode by.

"Kathleen, you ready for the cattle drive?" he said.

"Absolutely," she responded. "Bring 'em on."

She had no idea he was serious.

I had no idea *she* wasn't.

To all outward appearances, she was handling Painto just fine. But I learned later it was pure cover-up, an effort to not look wimpy in front of all the real Texas cowboys on the ride. Her fears had been rising all morning. And when fear begins to telegraph uncertainty to a horse, everything escalates.

It was not going well.

Then the trail boss brought everyone to a halt.

"Here's where we all split up and start rounding up those longhorns. Now listen up for your assignment."

"Rounding up what?" Kathleen asked.

Those words were barely out of her mouth when Jimbo got his assignment and loped off across the prairie. That turned Painto from a nervous, jiggy horse into a loose cannon.

Kathleen was done. It was all she could do to keep him in the vicinity.

"I think I should get off and walk back," she said.

"He'll probably be fine as soon as the other horse is out of sight," one of the riders advised.

But Kathleen was having none of it.

No longer was she experiencing the tiny fears about details that she was actually handling quite well. Now she had a truck-load-sized fear about all the things that *could* happen. *Might* happen. The kind of fear that, properly stoked, can become a self-fulfilling prophecy. Like on a ski slope when you are certain a particular hill isn't wide enough for you to make turns. If you believe it, it'll be true. And it was occurring to Kathleen that she might not be able to control a nervous wreck of a horse in the midst of an angry herd of stampeding longhorns.

In truth, this was definitely the kindergarten of cattle drives, barely an hour long, driving thirty or forty very calm longhorns from a pasture into Fort Griffin for the evening's festivities. I think there were at least twenty of us to do it. One rider for every two cows. *Not* the usual odds.

But to Kathleen, even one rider per cow wasn't enough. She had been arguing with Painto all morning, and now the vision of stampeding longhorns, all headed straight for her, was wedging itself into her imagination. She was beginning to lose it.

I would soon learn that Sonny Boy was also focused on a buddy. When Red and I rode out to take up our positions for the drive, Red and Sonny Boy's pasture buddy disappeared off into the trees. Sonny Boy wanted to follow. He got nervous and jiggy, and called out to his buddy, his fear and adrenaline rising. But a few quick exercises reminded him that he was okay with me. He'd jig right, and I'd say, *Let's turn a circle to the left.* When he'd try to go forward, I'd say, *Let's back up*, And every time he did as I asked, I would rub him on the neck, tell him "good boy," and put slack in my reins and legs to give him release and comfort. In other words, I gave him the choice. Be relaxed, pay attention, and remember that I'm your leader and you are safe with me, or plan on turning circles and backing up for the next hour. Get out of the reactive side of your brain and move to the thinking side.

It wasn't long before he forgot about his buddy and focused on a sprig of grass. He had a good leader right here who understood him, and things were just fine, thank you very much. We stood there, a herd of two, no one else in sight, for maybe ten minutes. When the time came for us to move forward toward the cattle, he was calm, focused, listening, and ready to do his job. Which, parenthetically, he knew way better than I did.

Kathleen had all these tools, but when fear is on the rise, it tends to take over. She was about to dismount when an older cowboy, a very generous sort, took her reins and said, "You can just pony along with me up the road until he loses sight of his friend, then he'll be okay. We'll stay well away from the longhorns and just chat."

Later she told me that as they walked down the road, the cowboy had carefully and calmly talked her down off the ledge until her fear was on the run. Interestingly, as generous as the old cowboy was, the thing that prodded Kathleen back to the job at hand was a statement he made about her relationship with the horse.

"I think you believe you can be buddies with this horse."

The old cowboy paused. Of course she believed that.

"You can't be buddies with a horse," he said flatly. "All he wants to do is get back to the herd. He cares nothing about you, nor will he ever. He will tolerate you up on his back because you make him tolerate it. You give him no choice."

Quite suddenly Kathleen was back in control.

She knew that she should've established her leadership with Painto from the beginning. She should've found transportation to the remote camp that first evening, or early the next morning. For two days she had been nothing but a passenger, albeit a knowledgeable one. She had mistaken Painto's calm for relationship.

She did not debate the old cowboy because she knew he had never experienced the kind of relationship with his horse that she had experienced with hers. He didn't understand that when you speak the horse's language, and give him choice, you *become* the herd and he doesn't need to get back to anyone. You are his herd and he's already there. You are the trusted leader. But Kathleen realized that to the cowboy, such a notion was as implausible as a Marriott on the moon. And she wasn't about to end the trail ride proving him right.

She had a word with Painto, running through a number of exercises to engage his intellect and establish at least some degree of leadership: circling, backing up, moving his hindquarters left and right. Ultimately bringing both horse and rider back to the thinking sides of their brains. Then she took sight on a longhorn, albeit a small one, and off they went.

Two horses.

Both with separation anxiety.

I had spent time with Sonny Boy. I'd gotten to know him, and him me. We had dug around in his thinking side, and he had realized that I could be trusted and respected. *Just remind me every once in a while,* he said, *and I'll be fine.*

Kathleen had done none of that, and when Painto's pasture partner disappeared, he needed a leader and didn't have one.

She had almost sixteen hours in the saddle during those three days, which helped her immensely. But the last couple of hours might well have been the most important, as she rose to the occasion of establishing leadership, and a true relationship. She and Painto definitely ended the trail ride as buddies. The best kind. We were both blinking back tears when we had to tell Painto and Sonny Boy good-bye.

Relationship makes a difference.

Leadership makes a difference.

Even with borrowed horses. Or rented trail horses, who carry folks around every day of their lives. You never know when it will come in handy for the horse to think of you as a leader.

And it's so much nicer to know that you're off on a ride with a friend. A partner who trusts you. Not some vacant-eyed mechanical device manufactured just to carry you around.

The rub, of course, is that leadership isn't easy or free.

With horses or in life.

It's earned.

But it *does* make a difference, and is worth every ounce of the effort.

21

Confined

The golden stallion stood on a small rocky knoll gazing at the canyon's entrance. He sniffed the westerly breeze blowing lightly at his back, then turned a full circle, scanning the entire horizon. There seemed to be nothing to fear, but his senses wouldn't relax.

Inside, through the outcroppings of the passageway, the canyon was green with grasses fed by a crystal clear stream that disappeared underground before departing the small gorge's towering walls. This was grass and water much needed by the herd, and especially by the matriarch, who needed nourishment and rest for herself and the foal she carried.

But once inside, there was only one way out. The stallion's herd had been trapped there once before by humans on horses and had lost several of their own. It was his job to make sure it didn't happen again.

The stallion would've turned back, but the matriarch was tired, absent her usual spirit, and that convinced him to have a closer look. The uneasy herd was squirming and pawing the ground. They could smell the fresh grasses and wanted to go in. All had come too far without forage.

The stallion sniffed the air again, then eased down the slope of the knoll and walked cautiously toward the outcroppings that formed the entrance. The herd followed, but the matriarch spun with pinned ears and stopped them in their tracks.

Minutes passed.

The matriarch was growing impatient and began to paw the ground. She needed sleep. Finally, the stallion appeared and stood aside, allowing the matriarch and the herd to pass into the canyon.

The grasses were rich and full of sustenance. A good meal and sleep refreshed the herd. A few of the yearlings cavorted and kicked up their heels. The matriarch had much needed REM sleep and clearly felt better, her spirit replenished. She was ready to move on, but when she turned to the entrance, her blood chilled and her terrified scream spun the stallion in place.

There in the passageway, silhouetted against the afternoon sun, was a human. A man.

Terror raced through the herd like lightning through a thunderstorm. The stallion screeched and reared, striking out with fear. Images of his sister thrashing on the ground flashed through his memory, her legs bound by man's lariat, her screams for help bouncing off the canyon walls. The great stallion's front feet hit the ground at full stride, racing straight toward the man who was blocking their way out.

The man was not on a horse. He stood alone between the out-croppings with nothing but a bedroll under his arm. The muscular stallion slid to a stop mere yards away and rose to the sky, pawing and prancing, teeth bared, snorting and screeching.

The man stepped backward, seemingly mesmerized by the huge pawing stallion, then gathered himself and moved slowly aside, to the top of a large boulder, where he sat with his back partially to the big horse.

The stallion was not sure what to make of this, but he called to the herd and they raced through the entrance, their protector standing bravely between his charges and the man. When all had disappeared outside the canyon walls, the stallion turned to the man on the rock, still pawing and snorting, but also curious. This man was not like any human he had seen before. He did not look the great horse in the eyes, but kept his head lowered, toward the stallion's feet. And made no attempt to approach him.

After a moment, the great palomino turned and raced off after his herd.

This venture into the canyon had ended well, but he would not go there again. He would not allow his herd back into confine-ment. They would have to find forage elsewhere.

22

Cute Hitching Posts

I remember it well.

My first encounter with a hitching post.

It was a Saturday-afternoon matinee at the local movie theater in Little Rock, Arkansas. I was six years old.

Roy came galloping into Tombstone on his beloved Trigger, scattering townspeople left and right. Trigger slid to a dazzling stop in front of the local saloon precisely as Roy's first boot hit the ground. One loop of the reins over the hitching post was plenty for this brilliant steed. Roy then checked both his pistols, spinning them on his trigger fingers at least twice before they landed perfectly, and simultaneously, back in their respective holsters. Then the King of the Cowboys strolled calmly though a pair of swinging doors into the boisterous smoke-filled saloon, accompanied by the tinkling sounds of an old upright piano.

These weekend adventures were always the same. Inside the saloon, the noise would dissolve into silence at the sight of the tall silhouette in the doorway, itchy fingers only inches from the carved-pearl handles of his matching revolvers. Folks would creep quietly away from the half a dozen filthy, bearded, drunken villains crammed around a poker table in the back. No one wanted to get caught in the cross fire.

Roy was outnumbered six to one, but he would walk slowly, spurs jingling with every step, toward the table in the back. The villains would begin to spread, obviously planning to get the drop on him. Things did not look good . . . but I was never worried about Roy. I was worried about Trigger.

Even then.

I remember vividly wanting to know what was happening outside. Poor Trigger was tied—okay, looped—to the hitching post, standing, waiting, with nothing to do, nowhere to go, nothing to eat. Just standing. Or worse, perhaps being stolen. After all, he wasn't even tied with a good knot.

Years later (a *lot* of years later) I was telling Kathleen this story one night while we were doing the dishes.

"No one would dare steal Roy Rogers's horse," she said. "They'd be afraid for their lives."

"This was no ordinary horse," I argued. "And there he was just standing at this hitching post while his leader was inside for however long it took to first negotiate with then kill half a dozen notorious bad guys."

"What's your point?" she asked.

My point was that, as a kid, standing in one place for long periods of time would've been no fun for me. So it stood to reason that it would be worse for a horse. Horses like to move around pretty much all the time. And they like to be with their herd, not standing alone at a hitching post. Come to think of it, I even wondered why the saloon was always full, but Roy's horse was the only one tied out front.

The whole concept seemed harsh to me. Tie your horse up and leave him as long as you wanted whenever it suited you. Never mind what the horse thought. Or felt. I totally bought into Roy's love for Trigger. Lock, stock, and barrel. So he wasn't supposed to treat him that way. He was supposed to care more. And back then I didn't even know that hitching posts like these could breed some pretty nasty accidents.

My first real-life experience with a hitching post was at a riding instructor's place where the family was taking riding lessons, just after we had acquired our first three horses. Seems backward, doesn't it?

I led the horse Kathleen would ride from his corral over to the hitching post where he would be tied while being brushed and saddled. It was a standard hitching post, just like the two we had installed at our place. You know, the cute kind, with two upright posts and one horizontal post that stuck out maybe a foot on either side of the uprights, like those in front of every saloon in every western movie ever made. But in the real world there seems to be a design flaw with this type of hitching post.

"Watch him closely while he's tied there," the instructor said. "If he gets a loop or two wrapped around one of the end pieces, the lead rope will get too short and he'll freak out."

"Gotcha," I said confidently, wondering, *If that's the case, why are the end pieces even there?*

I had no sooner gotten the thought out of my head when the scenario unfolded exactly as the instructor had described it.

The horse leaned across the horizontal post to sniff something on the other side and in the effort managed to loop the rope over the open end of the hitching-post rail. Twice. Suddenly he realized he was down to about two feet of rope and went instantly nuts. He pulled back so hard that he jerked the halter clip right off the rope. Thankfully he was an older, well-trained horse and as soon as he realized that he was no longer confined by the two-foot rope, he stopped, snorted, and just stood quietly, waiting to be retrieved.

That could've been my very first learning experience in the horse world.

Could've been.

The very next week I tied Cash to one of *our* cute hitching posts, with the ends sticking out, just like the one at the instructor's. And believe it or not, he did exactly the same as the horse at the instructor's place.

Almost exactly.

He didn't pull the clip off the lead rope. He jerked the entire hitching post out of the ground, concrete and all! This was one scary moment. Cash was dragging a seventy-pound hitching post around the yard, scared to death and wide-eyed crazy. And I was the same. I had only read about keeping my adrenaline down. I had not so much as even practiced it. I finally managed to get a hand on the rope and somehow calmed him before he seriously hurt himself, or me, with his dangling anchor. I was so terrified,

I don't even remember exactly what I did. But I do remember what I did next.

I went straight to the tool room, scooped up the Skil saw, and cut off the ends on both of our *cute* hitching posts.

No longer cute.

But infinitely safer. Such an incident could never occur again.

So many horse owners I've spoken with have recalled the same type of accident.

Why, I wondered, *doesn't everyone just cut off the ends?*

When I ask, the responses are mind-boggling. Everything from *Oh no, the hitching post is so cute this way* to *It's not that much of a problem; just happens every once in a while.*

Once is enough!

Highly respected vet and author Dr. Robert M. Miller says, "Horses categorize every learned experience in life as something not to fear and, hence, to ignore; or something to fear and, hence, to flee. This is extremely useful in the wild and utilizes the species' phenomenal memory, but it often creates problems in domestic situations. It is incumbent upon those who must work with horses not to cause bad experiences that the horse will forever regard as a reason to flee."

This is where our journey really began.

The hitching post.

The *cute* hitching post.

This was the first time we had questioned traditional wisdom, when applied against a rule of logic, and traditional wisdom had come up wanting. The first time I had come face-to-face with something that made absolutely no sense whatsoever. Something that could very easily be fixed to make the horses' life better.

Something that would cost no money and take very little effort. I talked to dozens of folks about it, including the riding instructor. But, to my knowledge, not one end has been cut off a hitching post.

They're all still cute.

And this was just the beginning.

Dr. Miller adds, "Horses have the fastest response time of any common domestic animal. Prey species must have a faster response time than a predator or they get eaten. We (humans) commonly interpret the flight reaction as stupidity."

It was becoming clear to me that we were not dealing with stupidity. Confinement without escape leaves the horse no choice but to react. To attempt to flee. Why does it make any sense to risk letting the horse categorize a tying experience as something to fear? It's up to us to make sure that he has no reason to fear it. It's up to us to cut the ends off the hitching posts.

Or, better yet, use a tie ring.

Remember the instructor's horse? Once he was free, no longer confined, his adrenaline went down, and he stood still. Just waited. I've noted this type of behavior so many times. When the horse is thinking, everything he's learned in the past is available. His memory is truly incredible. But when he's on the reactive side of the brain, thinking is trumped by the need to flee.

The goal, then, becomes obvious.

So how, I wondered, could I teach Cash that tying did not equal confinement?

It sounded like a pretty stupid question to me. Tying *is* confinement.

Yet there are times when a horse needs to be tied. For hoof work, grooming, his own safety.

Or at least given the illusion of being tied.

Hmm. *Illusion.* Good word.

I earned spending money as a magician when I was in high school, and I'm still fascinated with what people will believe when they want to. Maybe horses would do the same. I hit the Internet.

There are all sorts of complicated techniques out there designed to cure a horse from what are commonly called pull-back problems. Fear of being tied. Usually embedded by some incident like Cash's. Some of the techniques I found were so complex, they bordered on the ridiculous. Others didn't really address the issue, or just nibbled around the edges.

Then I stumbled onto the Blocker Tie Ring on Clinton Anderson's website. The simplest, least expensive, most effective little device invented since sliced bread. And its effectiveness relied on *illusion.*

Yes!

One ring costs about twenty bucks. And solves pull-back problems forever, quickly and simply, because it addresses the issue from the horse's viewpoint. Why everyone in the entire horse world doesn't own a dozen of these is beyond me. Of all the folks I've shown it to, only one person had ever seen one before.

It's a small stainless-steel ring with a pivoting tongue across the middle. One loop of the lead rope around the tongue, and that's it. No knot. The tongue applies enough resistance to the lead rope to give the horse the sense of being tied, but if something scares the horse enough to make him pull back hard, he gets relief, not confinement. The rope slips through the ring just enough for the horse to realize: *Hey, I'm not confined. It's okay.* Just enough to send him back to the thinking side of his brain.

Cash was afraid of being tied from the moment he uprooted our hitching post. The tie ring solved this problem in less than thirty minutes. He learned very quickly that tying no longer meant confinement. Following Clinton Anderson's model, I would hold one end of a long lead rope with Cash on the other, and the rope in the tie ring. Then I'd run at him yelling and waving a plastic bag in the air. Certain that I had morphed into some sort of banshee, he at first freaked out and pulled away. He leaped backward several steps, with the rope sliding through the ring, maybe six or eight feet. The flight reaction. Flee first, ask questions later. But realizing he wasn't confined, and that the banshee was actually me acting like an idiot, his adrenaline would drop, he'd eventually stop, and I'd praise him and rub him with the plastic bag. Back to the thinking side he'd go.

Then we'd do it again. And again. Soon he was no longer moving at all, not pulling back so much as an ounce. In fact, he began to give me that cocked-head questioning look of his as if to say, *Why in the world are you acting so silly?* He had learned very quickly that he was not confined and could move away if he had to, and that neither Joe nor the plastic bag was going to hurt him.

No more worries.

Except that Kathleen was afraid the neighbors were going to call the police about the crazy screaming guy up on the hill.

Dr. Miller says that horses not only respond quickly to flight stimuli but also are genetically disposed to return to normal quickly when they realize that what they thought was frightening is actually harmless. "If this weren't so," he says, "in the wild, they would spend all their time running, and there would be no time to eat, drink, rest, or reproduce."

Watch them in the pasture when the sudden flight of a bird or a blowing piece of paper startles them. In an instant, they react. Flee. But, then, in two or three "instants" the adrenaline drops, they realize they are still safe, and it's back to munching.

Just a few days ago, Cash was standing near the fence nibbling on a weed. Suddenly he reared and performed a perfect 180-degree rollback. In a flash! Something had ignited the flight reaction. But before his front feet hit the ground, he was already back to the thinking side. *Oops. Not a predator. At least I don't think it was a predator.*

He turned to the fence and stretched to see over a boulder. Needing a better view, he took a few steps to his left, then stretched to see *around* the boulder. It was funny to watch. His ears were up and his expression was full of curiosity. Not fear.

I walked over.

"What do you see, Cash?"

I followed his gaze. It was a rabbit, maybe ten feet away. Frozen in place.

"Ah, it's just a rabbit," I said.

But because the rabbit wasn't moving, I don't think Cash was convinced. He seemed to be waiting for confirmation.

"Here, watch," I said.

Cash watched me pick up a small rock. I pointed toward the rabbit and he looked in that direction. I tossed the rock in the general vicinity of the rabbit, and when it hit the ground, the bunny scampered off across the dirt. Cash stretched even farther to watch it disappear down the hill. Then his attention returned to the weed he had been nibbling.

It was a fascinating display of how quickly a horse can react, and how rapidly he can also come back down. Back to thinking. Even curiosity. How quickly he can desensitize himself to something that only a moment before had created a flight response.

This was interesting to witness and was logged in for the future.

We now have tie rings all over the place. And a couple in a tote bag to hang up when we're away from home. And now if Cash, or any other of our horses, needs to be actually tied to a post or a rail—so long as it's not a cute hitching post that could cause an accident—there are no issues at all. Tying is no longer associated with confinement.

Well, except for Skeeter. The first time he was on a ring, we stepped into the tack room for a moment, and when we returned, he was up in the yard munching away on grass. He now gets two loops around the tongue of the ring. Enough resistance to keep him handy, but if something were to really freak him, he'd still get relief. Note that his departure from the tie ring did not involve flipping to the reactive side of his brain. He was definitely on the thinking side, using his eighteen years of experience to realize, *Hmm . . . I'm not really tied here, and that grass is awfully green.*

Did somebody say horses can't think rationally?

I was telling a trainer about all this and he said, "You don't need to go spending all that money for rings. Just saddle up your horse with the heaviest saddle in the barn, tie him to a rock-solid post, and leave him there all day. He'll get the picture."

No jest. He actually said that.

I prefer the rings. And so do our horses.

One more trust issue put to rest by approaching the problem from the horse's end of the rope. By understanding why he was afraid in the first place and addressing that issue. It took very little effort to dig out that knowledge. And even less to discover the tie ring. A shiny, bright, inexpensive, simple device that quickly and easily solves an age-old and complex issue with horses that no one I know in our community could've told me about.

Why?

Because they didn't know about it. We had to find it on our own. We had to figure it out.

Every time, without exception, that I've attempted to take the easy way out of something, to just go the traditional route, or let things slide, or let someone else make my decisions, I've always regretted it. Because it's never the easy way out that makes things happen, that makes life better, that gets the job done.

Dr. Kenneth McFarland, a noted motivational speaker, often told a story about driving up a Vermont mountain road and coming up behind a huge flatbed truck loaded with hundreds of big crates with little holes in them. The truck was just barely making it up the hill. It was a narrow, winding road, and Dr. McFarland couldn't get by, so he settled in behind . . . and before long, the truck stopped and the driver got out with a short two-by-four, walked around to the back, and began beating the truck.

The doctor's mouth dropped open. *Whatever could this guy be doing?*

A few more miles up the mountain, the truck stopped again, and here came the driver again with his two-by-four, beating on the truck.

Well, the good doctor stood it just about as long as he could, and the next time the truck stopped, he got out and approached the driver.

"Pardon me, sir. I couldn't help noticing how you've been

stopping every few miles and beating up on your truck there, and I must admit my curiosity's got the best of me. How come you're doing it?"

The driver said, "Mister, I've got me a two-ton truck here, and I'm hauling four tons of canaries. Unless,I keep half of 'em flying, I'm overloaded!"

That man definitely did not take the easy way out of his predicament, but he was getting the job done!

In my life, it's up to me to get the job done. To make things happen. When life deals me four tons of canaries, and I've only got a two-ton truck, it's up to *me* to keep those birds airborne! Not somebody else.

Kathleen and I made the decision to care for these horses. They made the choice to accept us as leaders. We have no alternative but to always strive to make our decisions based upon our own knowledge, not someone else's. To keep asking questions. To keep challenging traditional wisdom when it defies logic. To spend the energy to gather that knowledge and make every attempt to understand it.

I've simply never found a better way.

It's true that Kathleen and I haven't been at this horse thing very long, but I can promise you that our horses recommend this approach to problem solving.

And shouldn't that be what it's all about?

23

The Legacy of Sojourn

Assume the predator stance. Swell up like a balloon. Eyes wide. Eyebrows up. Be a horse. A leader horse.

Now shake that lead rope and wag your finger at the same time.

"Back up! Back up!"

Some say that horses are not verbal so it's best not to give them verbal orders. Restrict the teaching to body language.

I say use everything you've got, especially in the beginning.

Shake the lead rope. Wag the finger.

"Back up!"

Sojourn took a step backward and I dropped the pressure

immediately. Release from pressure equals reward. Reward equals learning.

This was one scary horse. Not because he wanted to be. Not because he was mean. Because he was *big*! And he was very smart. And I was very new to all this.

And, unlike any other horse we'd met, he was very possessive. He wanted *all* of my time. *All* the time. But he was only one of four, soon to be one of six. I was beginning to wonder if Join-Up had worked too well? Was the bond too strong?

Because he was so bright, he became bored easily and would destroy virtually anything in his stall just to have something to do. This was before we discovered that stalls and horses were an unhealthy mix. Sojourn was trying to tell us something.

His relationship with the other horses also left much to be desired. He would become upset when anyone else received attention. He's the one I allowed Cash to be stalled with, which resulted in a gashing kick to Cash's forehead. It was becoming a problem.

Still I persisted, working on relationship, doing lots of groundwork. *Back up. Move your hindquarters. Come in. Go this way. Now the other way.* He listened well and began to learn, and was quite willing to stay in the arena all day. *Never mind those other nags,* I could almost hear him saying. *I'm your man.*

Very soon it became apparent that no one else in the family wanted anything to do with Sojourn. Other than to rub him . . . from across the fence. I was the only one who would work with him, and, frankly, I rather enjoyed seeing the progress with such a high-strung, possessive horse. But he was keeping me away from Cash and the other horses.

He was one of our first two horses, and a perfect example of what *not* to do when looking for a horse. He was young, gorgeous, very athletic, and would hang with you like a puppy.

Awww, he loves me already. Look, he wants to be my buddy.

This was before I knew that *buddy* was defined differently within the herd. Before I had learned how valuable it can be to spend time with the horses, just observing the language, the interaction. Before I had discovered that a strong and effective relationship with a horse required, in effect, the human to become one. This was back in the beginning, and I had a lot to learn.

A lot.

This was also before I knew that I could search deeply into a horse's eyes and actually feel what was going on inside. Is this a kind horse? A gentle horse? Is it a willing horse? Will he accept a leader easily? All of that was yet to come. With experience. And lots of time in the pasture. As was the knowledge that we couldn't rely on sellers to tell us what we were not experienced enough to see.

But I did know how to get this horse onto the thinking side of his brain. Give him a task, or stir his curiosity. Divert him from the reactive side. And keep the adrenaline down. His, and mine.

Sojourn was smart, but not a particularly quick learner. He didn't seem to want to be bothered. I believe he figured he could take fine care of himself, thank you, and what was the point of all this anyway? Maintaining a leadership position with him was eating up all of my time. What I didn't know then is that he was teaching me a very valuable lesson.

It wasn't long before Kathleen and I began to nibble at the edges of putting Sojourn up for sale.

"I don't think so," I said. "He's gonna be a great horse. He has a lot of issues, but I can do this. I can bring him along."

"At what price to the other horses? At what price to your own learning curve?"

It was a conundrum. I actually liked him a lot. But he and I didn't have the connection that Cash and I had. It was just different, and difficult to explain.

I wanted to be with Cash.

I enjoyed being with Cash.

I wanted to *teach* Sojourn. The primary enjoyment from the relationship was seeing thresholds give way to progress. I suppose if he hadn't demanded so much of my time, it could've been different. But the mission was becoming a chore and really began to wear on me.

"He needs someone who only wants one horse, one focus," Kathleen admonished one morning over cappuccino. We were watching the horses pace back and forth in their cute little stalls. "Someone who will fill his day and spend time only with him."

"I know. I know. But that would be like giving up. Like failing."

Another scorpion.

I do hate to fail.

Mistakes, unfortunately, are a natural consequence of doing. The only way to avoid mistakes is to do nothing. When you're trying new things, taking risks, pushing for perfection, moving the ball forward, you're going to take some hits. The challenge is to go ahead and take the hits, admit the mistake, swallow your pride, use the error as a learning experience, climb back onto your feet, and move forward.

"I can't sell him," I said.

Kathleen sighed.

"You're becoming obsessive about this. You know that, right?"

Of course I knew it. But I didn't nod. I think *pout* would've been the operative word.

"So what if you do cause Sojourn to be the lightest, most receptive, most responsive horse in the state, what then? You don't care anything about competing. And you're never going to leave Cash. You enjoy him too much. So what then? You'll either have to keep up the training to prevent him from slipping backward, or you'll sell him to someone who'll appreciate him. Why not do that now?"

I studied on that for some time. "Because he's so big he scares me," I finally admitted. "I'm afraid to ride him, and I need to get over that."

That wasn't easy. Males aren't supposed to admit fear. I knew Kathleen had fears, but until that moment, she had no clue that I did as well.

"I suspect you're afraid because you don't believe your riding skills are good enough for Sojourn yet. And that makes him the wrong horse to learn on. I would think you should learn on a horse that makes you feel comfortable. You cannot concentrate on teaching the horse while fear is running out your ears."

Where have I heard that before? Or rather, where would I hear it again, months later? Foreshadowing.

"Don't hammer me with logic," I said. "It's not fair."

Kathleen smiled.

The decision was made, but I can't say that I was ever 100 percent in favor of it. Consent was all tangled up in my feelings for Sojourn, along with a smidgen of ego, the notion of failure, and the fact that we had no idea where this journey was taking us. By this point, I was reasonably certain that something was up. That God was leading us down this trail for a reason. But, unfortunately, God has never felt obliged to keep us apprised of His intentions.

Seven months after bringing Sojourn to our place, we decided to deliver him to our benevolent horse broker friend. Horses are her business. She is very picky and always makes sure the match between buyer and horse is a good one. If she didn't like the buyer, she would not recommend a sale. This was the same woman who had presented Cash to us. We told her that Sojourn must go to a good, kind home with an owner for whom Sojourn would be the only horse. Not a one-horse household. But a one-person horse. It saddens me to note that he still wore metal shoes because we had not yet reached the barefoot junction of our journey. I promised Kathleen to send Sojourn's new owners a copy of this book.

Meanwhile, the decision to sell fueled other problems.

We had to move him over to his interim home. And that meant we finally had to use our trailer.

The trailer we had raced out to buy one short month after the first three horses came to live with us.

"Why do we need a trailer now?" Kathleen had pleaded. "We don't even know what we're doing with these horses yet."

"You never know," I said. "You never know when we might need it."

I thought at the time that perhaps the scariest moment of my life was pulling our new gooseneck three-horse-plus-tack-room trailer back home from the dealer. It wasn't. The scariest moment was getting this twenty-nine-foot monstrosity up our driveway, a three-hundred-foot vertical rise from top to bottom that seems to go straight up in spots. One of the steepest points is a hairpin turn, at least 270 degrees, through a gate!

For reasons that carry no more logic than my original belief that horses should wear shoes, it had never occurred to me that the trailer might not make it up the driveway, through the gate, and around the turn. That thought hit me about halfway home. Suddenly, pulling the trailer down a traffic-logged California freeway was no longer an issue. I was terrified that I would get halfway up the driveway and that would be it. Our new used 2001 Dodge 2500 pickup would just quit. It was not four-wheel drive. Why? Don't even ask. For inexplicable reasons, my car is, but the pickup acquired to pull this huge trailer isn't.

If we had to stop for the gate to open, would we be able to start up again? And could we make the hairpin turn, which would have to be done very slowly to be certain the trailer didn't trash the gate trying to squeeze through? And if it *did* go through and made the turn, would it continue up the steepest part?

The good news is that the truck performed valiantly and the trailer missed the gate with at least two inches to spare. The bad news is that the trailer sat at the top of the hill for seven months before it ever moved again. I suspect I was more afraid of the trailer and the driveway than I was of riding Sojourn.

"So tell me again why we needed this trailer way back in June?" Kathleen asked often.

"Er . . . uh . . . practice?" I would squeak.

Practice was actually a pretty good answer. It didn't address the reason why the trailer hadn't moved in half a year, but having it had enabled us to learn a new skill. Before purchasing the trailer, we had never loaded a horse, *any* horse, into any trailer.

All of our horses responded differently to the experience.

Some acted as if they had never seen a trailer before. Others—Cash—walked right in and started chomping hay.

With Join-Up, and Monty Roberts's trailering techniques, what apparently can often be a very stressful time for man and horse went very well for us. When the horse trusts and respects you as his leader, he is willing to try uncomfortable tasks. In the simplest form of trailering technique, after the horse is taught to back up, Monty teases him with the trailer, walking him up to the door, and then backing him off. *Nope, we don't want to go in that big old nasty trailer yet. No sir, we won't make you do that. But let's go up and have another look.*

The horse is walked back up to the door again, given a moment to sniff and look, then backed away again. Over and over this happens. Again and again. Until finally you can actually feel the horse *wanting* to go in. *No more backups, please!* At that moment, Monty keeps walking right into the trailer and most horses follow him right in. Those who don't go back to the beginning and do some more backing up. It's fascinating to watch, but even more fascinating to see how well it worked with our own horses.

All six of ours wound up going into the trailer quite willingly, happily, within a few hours at the most. Not one was ever forced, even a little. The choice was always theirs. Even with Sojourn.

Pocket gave us one of my favorite memories of these lessons. The first time I stepped into the trailer asking her to follow, she stopped short, looked at me for a moment, and stood straight up on her hind legs! A full-blown Roy Rogers and Trigger kind of rear. No pawing the air. No intent to harm. Just saying, *I don't want to do that just yet.*

I stood there with a silly grin spread across my face, less

than ten feet away from this big paint horse who was suddenly about twelve feet tall. *This is going to take a while*, I thought. But I couldn't have been more wrong. I backed her away once, then walked up and stepped into the trailer, and she practically ran over me coming in behind. No resistance whatsoever. Go figure.

The day we loaded Sojourn to move him, the emotions were not as happy, and I'm sure he was reading our mood because he was clearly a bit tense. He walked right into the trailer and I closed the partition. But when I closed the trailer door with no other horses in the trailer, he became seriously fidgety and pawed the floor. As we drove out right past the other horses and his one herd buddy, I could hear him stomping around and calling to his herd mates, which did nothing to settle my nerves, already scrambled from the decision to sell, and from navigating my first trip down the driveway with an actual passenger in the trailer.

It was less than a fifteen-minute ride, but I was very relieved when we arrived. As I opened the door, Sojourn became more agitated. To say I was nervous would be an understatement. This would be a test of will to keep my adrenaline down. The trailer looked much smaller now than it had in the past. When I opened the partition, it would be just me and that gigantic horse in this tiny little trailer. A gigantic horse who was clearly not very happy about something. I had no idea about what until the partition swung open.

His front right foot was caught in the hay bag, about chest high.

He was terrified. He couldn't get away from it. When the partition swung open, he wanted to bolt, but couldn't. He was tugging and stomping, and trying to rear enough to get his foot out. Should I run? Get out of harm's way? Should I try to calm him and get close enough to unsnap the feed bag? That option didn't seem very intelligent at the moment. I swelled up like Jabba the Hutt, shook the lead rope, and wagged my finger in his face.

"Back up," I said, as if he could. "Easy."

I was trying to control my own adrenaline. All of this happened so furiously fast that I have no idea how or why I decided to throw caution to the wind in favor of trying to make him focus on a task, get out of the reactive side of his brain, and, hopefully, get calm enough for me to approach his foot and unsnap the feed bag.

Outside the trailer, Kathleen was frozen in place, afraid to even think of looking inside. She was certain from the stomping she heard that I was being trampled to death. At some level I suppose I had considered that possibility but had decided that I would attempt to rise above it. Do whatever had to be done and do it well enough to make it work. To keep both myself and Sojourn safe from injury. I'd worry about what could've happened later.

Inside, Sojourn's eyes were as big as saucers. All I could see was white, but he was watching me. He seemed to be listening. Or trying to. Then suddenly the feed bag tore loose and his foot was free. This could be good, or it could be terrible. I shook the lead rope harder, wagging my finger at light speed.

"Easy. Good boy. Back up. Just a bit. Pay attention to me."

My adrenaline wanted to soar, but amazingly didn't. I could *feel* the calm, the lid staying on. I knew this was the answer. If I was calm, I had a shot at causing Sojourn to be calm. He was blowing and snorting, his eyes still crazed, but he was standing still, more or less. He was listening to me. I eased toward the door, consciously deciding whether to toss the rope and let him bolt or step in front of him and rely on my ability to keep him calm. Relatively speaking.

I chose the latter. I stopped him. Backed him up a step, which he accepted. Then I slid in front of him and stepped cleanly out the door and immediately to my right, out of his way. He came right out behind me and loped off to the end of the lead rope. I signaled him to come in rather than walking to him, wanting to keep him focused, thinking. He huffed and puffed and snorted, but walked slowly up to me and I rubbed his forehead.

"Good boy," I said. "*Very* good boy."

I had never meant anything more in my life.

I was so proud of him. And of myself. And of the fact that this event, as traumatic as it was for Sojourn, me, and Kathleen, would forever live as the certain proof that all we believed and all we were doing was right and good. That when a horse has accepted us by his free choice and has assigned us the position of herd member and leader, and when we accept and live up to that position, he will defer. He will subordinate even his worst fears to the trust he has placed in us.

When Kathleen finally tiptoed up to us, tears were running down my cheeks.

"Did you see that?" I blubbered.

"I did."

"Did you see how good he was?"

"I did."

She rubbed him on the forehead.

"I can't leave him here," I said.

"You also can't do it all," she said. "And he wants it all."

She let that sink in for a moment.

"Our decision is a good one," she said. "He needs a full-time leader. You've done an amazing job with him, and he has taught you that the sky is truly the limit when you walk in the horses' footsteps, when they make the choice, and when you fulfill their need for trust and leadership. He proved the truth of that today in spades, and we can love him for it. But remember, part of loving, the hard part, is making sure that he has what he needs. Sojourn needs someone else. And our other horses need you."

It was a quiet ride home. For once I wasn't thinking about the big trailer that was following us.

"It's interesting," I said, "how a single thought or event can feel both good and bad, can cause hurt and yet empower."

"Yes it is," Kathleen said. "It is indeed."

Neither of us spoke again. We were thinking about Sojourn. And the legacy he had left us.

We now knew it was all for real. It was all true.

Good boy, Sojourn.

24

The Big Round Circus Ball

When I first walked Cash down to the arena to look at this new object, he stopped cold in his tracks and just gaped at it.

There was nothing in his memory banks that looked like this.

Then a breeze wafted by and the darn thing moved.

Omigod, it's alive!

The next thing I knew Cash was at the other end of the arena, huffing and puffing. No idea how he got there. Seriously. None.

"It's just a circus ball," I hollered.

The look he gave me spoke volumes. But this circus ball and Cash would soon change my mind about a lot of things I needed to relearn.

▸━◦━◦━◂

IMAGINE AN INDOOR arena. Twelve thousand wildly screaming fans, there to see the season opener of a brand-new professional arena football league. Music blaring. Drums pounding. Feet stomping. Spotlights undulating.

The perfect atmosphere for a horse, right?

I cannot imagine taking one of ours into that fray. And yet Hasan, a majestic gray Arabian stallion, would gallop down a smoke-filled tunnel and out into this chaos, running right through a large rubber blow-up football helmet. The racket would escalate. Fireworks would explode into light and thunder as Hasan galloped around the arena. He would then rear and execute a hind-leg walk with his front legs landing perfectly on a pedestal upon which he would pivot in a complete circle, saluting all the fans with his right front leg. He would then stand motionless while a rumbling procession of motorcycles roared out of the inflatable helmet and circled him, delivering cheerleaders in short skirts to the field.

Even more clamor.

And Hasan just stood there!

Motionless. Relaxed.

As if nothing whatsoever was happening!

I couldn't believe it.

And it all started with a circus ball.

Hasan's trainer and constant companion of seventeen years is Allen Pogue. Allen's work has cast him as a trick trainer (but the word *trick* does not even scratch the surface and seems to diminish the value to me). What he accomplishes for the horse is so much more than tricks. We've just recently discovered Allen, and we're dumbfounded by how his horses treat him, and try for him. And have fun doing it.

Fun is a key word here because once basic natural training of a horse has begun in earnest, after the horse has been given the choice of whether or not to be with you, the work is all about maintaining leadership and relationship. But the repetition can become boring for owner and horse. Allen Pogue's training of self-motivated behaviors is all about removing the boredom, engaging the brain, and having fun. And communication is no longer a one-way street because the horse can now do something on his own that will speak to you.

According to Allen, the typical ranch horse or performance horse does not do much reasoning because he's never asked to. So much typical training is based on the horse's genetic desire to be safe and comfortable that the usual learning process is heavily slanted toward giving the horse the choice of either doing the behavior or being uncomfortable. Like the simple request that the horse lower his head. It's either lower it or feel the discomfort of halter pressure on top of your head.

I'll take the comfortable route, thank you very much.

The horse learns. There's no pain or cruelty. But not much reasoning either. And not a whole a lot of fun.

Which is why Allen's methods are so amazing.

At no other time, other than perhaps during a frolic in the pasture, do we ever get to see the horse *having fun*. Especially while his brain is engaged and he's learning.

Fun?

What's that about?

Most folks grow up assuming that the horse's capacity to reason and his ability to have *fun* are just not part of his genetic makeup. And unfortunately those subjects just never come up.

Didn't with us.

Just as with horseshoes, we never really thought about it. I was so focused on becoming one of the herd, using their language, directing them away from the reactive side of their brain—all of which is absolutely necessary to establishing a positive relationship with the horse, and necessary for the horse's basic training for respect—that it just didn't occur to me that a horse could reason, much like a dog can reason. Or that the horse could develop a verbal vocabulary, like Benji. The caveat is that all the basic training must come first, because neither reasoning nor vocabulary will occur unless the horse trusts you enough to stay on the thinking side of his brain, and respects you enough to choose you as a herd leader. Without that there is no opportunity for communication in either language, his or ours.

But why, I scolded myself, especially after years of experience with Benji, did it never occur to me to use verbal vocabulary, or to expect the horse to be capable of rational thought. It was frustrating that none of this bubbled up until I began to worry that I was boring our horses with repetition.

Was this another episode of following the crowd?

"I keep telling myself I'm a logical thinker," I said to Kathleen one evening when we were discussing it. "But I'm beginning to wonder."

"I don't believe we've ever heard a trainer or clinician use the word *reasoning* in reference to a horse," she said. "Not until Allen. And most clinicians we've seen advise against using verbal cues."

Now I was beginning to wonder why.

"Perhaps because the horse's language in the herd is mostly visual," she said.

"So are dogs in a pack, but Benji understands a huge vocabulary."

"Why do you worry so? The timing is perfect. You just said that everything we've done had to go before trick training."

"It's not *trick* training," I said. "It's self-motivated behavior."

"That's ridiculous," she said.

I sulked off to the computer to read more about Allen. And to see if he sold circus balls. Something new for Cash to focus on. Variety. Something different.

Cash in particular needs this because he's so bright. I was teaching him to back through the arena gate, a fairly scary thing for most horses, because a horse's only visual blind spot lies directly behind him. In front, there's a small area, right between the eyes, where they cannot see when the object is very close— which is why it's significant when a horse allows you to rub him on his forehead. Allowing you to rub where he cannot see definitely means he trusts you. But directly behind him he can see nothing unless he turns around or swings his head to the rear. For that reason it has taken quite some time to train some of our horses to back through the open gate. It makes them really nervous. But, like rubbing on the forehead, the process is ultimately a good thing for them because once they finally relax and do it comfortably, they're telling me, *Okay, I trust you. I am no longer afraid that you will back me into a horse-eating fence.*

And the relationship takes a step up.

The second time I walked Cash up to the gate and began to move his butt around to back him through, he swung right around and backed through all by himself.

The *second* time!

Okay, I've got it. What next?

How about a circus ball?

Bring it on.

That's how I stumbled onto Allen Pogue's Red Horse Ranch.

And learned that horses can have fun.

And can grow to understand words. Even sentences.

And are fully capable of reasoning.

It was like the barefoot moment: another cold, wet rag in the face.

Another *duh*.

I've been encouraging people for years in talks and seminars to exercise their brains every day. Take 'em out for a jog. The brain, like any other part of the body, works better the more it's used. And the more it's used, the better it works.

And this amazing phenomenon is not exclusive to humans.

Yet another epiphany.

The horse is a flight animal. Engaging his brain could be even more important to his ability to focus and reason than ours. It helps him control his own reactive side. Like Hasan did in that boisterous, explosive indoor arena setting.

"Please, Mr. Camp. I can't print that. Animals can't reason."

It was a reporter for the *Dallas Morning News* doing a story about the filming of the original Benji movie.

I was astonished. I had just spent thirty minutes ranting giddily about the unique concept of a dog *acting*, about the incredible facial expressions Benji was giving us, about those big brown eyes and the reams of dialogue they were speaking, about the dog himself and how for the first time I had come to realize that the story we were telling wasn't purely the emotional petition I had once thought but, in reality, quite plausible. Dogs, I had discovered, *can* think rationally. Can reason. And this particular one was extraordinary.

Not that other dogs aren't. Or horses. Or birds.

But most animals who have the intelligence, attitude, and tempcrament to do what Benji was doing never have the opportunity to learn and to gain the vocabulary that Benji has.

"Vocabulary? That's ridiculous!"

I bit my tongue because we were on the air. This was later, during a radio talk show in Norfolk, Virginia.

But Norfolk radio notwithstanding, Benji does have a vocabulary. And now I was beginning to realize that Cash could have one as well. He could think, and he could understand concepts. Just like Benji. Concepts like *other*. If you ask Benji for a foot, then ask for the *other* foot, he switches. If he walks off toward a chair and is told to go to the *other* chair, he looks back to see *which* one, then takes the point and heads in that direction. He understands the concept of words like *slow, hurry, easy, go on,* and *not,* no matter how the words are applied. When he is asked to perform a difficult task, you can actually witness the process as he studies the situation to determine the best approach.

But none of this is particularly unusual. Sheepdogs in Europe tend entire flocks *by themselves* for months, keeping the sheep together, deciding when to move them from one pasture to another, even stopping the flock to check for vehicles before crossing a road.

I read about a horse who was taught to bring a small herd of cattle in from the pasture every week and put them in a pen for a screwworm checkup. He would do this religiously, completely on his own. After a few weeks he decided, again completely on his own, that it was quite a bit easier just to keep them in the pen than to have to go fetch them every week. So he did just that, refusing to let them out.

At a press conference in a Miami hotel suite, a dozen reporters watched Benji perform one of his standard show routines, completely unaware at the time that he had made a mess of it and would've never finished had he not been able to reason it through.

He was wedged between two banister poles, pulling a coffee mug tied to a string of leashes up to the mezzanine level, which overlooked the group below. A person, of course, would use two hands, one over the other, but Benji used his mouth and a foot. He would reach down and pull up a length, hold it tightly against the floor with his foot, then reach down again and pull up another length, hold it with his foot, and so on, until he had retrieved whatever was tied to the other end. As he performed on this particular day, the leash slipped over the corner of the mezzanine floor and, because he was so snugly wedged between the banister poles, he could no longer reach it with the foot he had always used to hold it. I marveled as I watched the wheels turn. He pondered the situation for only a few seconds before he, quite logically, placed the *other* foot on the rope—the foot he had *never* before used to hold it—and went on with the routine as if nothing had happened.

Benji even understands what he's doing when he's acting.

"Now you've heard it all, folks. The dog understands he's acting! I suppose he gets script approval!"

Chicago. Another talk show host.

One of the more important sequences in the original Benji movie involved Benji moping forlornly through town. He knew that children were in danger but was unable to communicate what he knew to the family, who had, in fact, scolded him for trying. For the sequence to work, indeed for the entire *story* to work, these scenes had to generate unencumbered empathy and support for Benji's plight. He had to look as if he had lost his last friend. His desperation had to reach out from those big brown eyes and squeeze passionately upon the hearts of the audience.

It worked so well that during the first rehearsal, I almost aborted the sequence. I was forty feet above the scene with the cameraman and camera in the bucket of a cherry picker—the kind utility companies use to fix power lines—and Frank Inn, Benji's trainer, was in the alley below *screaming* at Benji, "Shame on you! Put your head down! Shame, shame on you!"

Benji looked as if he had, *in fact*, lost his last friend. It was perfect. I *believed* him. But I couldn't bear to see him hurt so from the scolding.

I asked to be lowered back to the ground and I walked into the scene and asked Frank to hold for a minute while we talked.

"What's the matter?" he asked, eyes wide and curious. "Isn't this the look you want?"

"It's perfect," I said. "But I don't feel right about getting it this way."

"What the hell are you talking about?"

"I don't feel right about you scolding Benji like that."

Frank's eyes rolled heavenward. "Turn around," he said. "Does that look like a scolded dog?"

Benji was aimlessly scratching his ear. He looked up at me and yawned idly.

"Watch closely," said Frank. He motioned Benji onto his feet and began scolding him again. Our floppy-eared star's head dropped like a rock, his eyes drooped, and he looked as pitiful as anything I had ever seen. Then Frank relaxed, chirped a simple "Okay," and as if he had flipped an emotional switch, Benji blinked away the blues, had a good shake, wagged his tail, and awaited his next command. He fully understood what was going on, and scolding wasn't it. He might not have known the word, but he was, in the truest sense, acting.

He picked things up so quickly that he even astonished Frank on occasion. Like the time we realized he had deciphered what the word *cut* meant. We were all on the floor, crunched around the camera, down at Benji's eye level. When the shot was over, Frank began to unravel from the pile and suddenly realized his dog was nowhere in sight. "Your dog's no fool," one of the crew chuckled. "When Joe said 'Cut,' he split for the air-conditioning."

Benji learned very quickly that the air conditioner was *his*. It was used to keep him from panting, so that's where he was supposed to be when the camera wasn't rolling.

But telling these stories, and a dozen others like them, left not the slightest dent in the armor of the *Dallas Morning News* reporter covering the newly emerging film production scene in north Texas. The story came out the next day on the front page. It was all about a seven-year-old Dallas girl and a nine-year-old Dallas boy making their motion picture debuts. The dog was barely mentioned. And there was certainly nothing about his ability to think. Or to reason.

Animals can't do that.

Not dogs. Not horses.

But don't tell that to Allen Pogue. He won't believe you. And he'll probably tell you this story about his mare Hasana.

"Whenever it was time for our young horses to be introduced to Liberty training—running free in a round pen without a line or lead rope—Hasana would be put in the ring with them, and she would help keep them in their assigned places in the lineup much better than any human handler could possibly do. If one got out of his place, as Gater would often do, she would trot right up alongside of him and promptly push him back into the lineup. If he resisted, she would become more insistent and give him the bossy-mare look or a nip until he resigned himself to his job. The precision and understanding that she displayed in this responsibility was amazing and her ability to teach other horses was a considerable help to me."

Many trainers, if not most, do not agree with Allen on the use of treats. And, as mentioned, most eschew the use of verbal cues. But Allen uses words and treats to build brainpower, relationship, communication, and fun. And, as they say about truth, when you see the results, the value is self-evident.

Kathleen notwithstanding, I like to call Allen's methods self-motivated behaviors because the horse chooses, on his own, to do it, or not. And there's no discomfort if he chooses not to.

It's worth repeating that a good relationship based upon choice, trust, respect, and leadership needs to be in place before training with treats is introduced. Or the horse needs to be started very young, as Allen does now with all of his foals. Otherwise, many horses—and owners—can become treat crazy, which is good neither for the horse nor the owner. Once a horse is listening, his brain engaged, a bit of treat here and there, given at just the right time, can actually encourage listening and reasoning.

Wouldn't it to be nice to walk into a pasture of six horses, call out to a horse by name, and have one horse look up? The right horse?

Or be able to say, "Lower your head, please." And have it happen.

Or have a horse come to greet you with a big, openmouthed smile? That's the first of Allen's behaviors that I began teaching Cash. Actually it looks more like a tooth-and-gum—showing pucker to me. As if the horse were offering a big, sloppy kiss.

Horses use their soft, floppy lips to sort through things, to check things out, to groom, to push away the dirt on a single strand of hay, or to tickle open your fingers to get to a treat. Which is the way the cued *smile* begins. As Cash's lips tickle my fingers, I raise my hand into the air, flip up my index finger, and say, "Smile." Cash had this down in no time. Even I was amazed. Now I need only to raise my hand and flick the index finger a couple of times and his lips open wide in an exaggerated goldfish pucker. And he holds it until I lower my hand.

He even offers the smile, on occasion, without my asking. Instead of exhibiting the typical pushy, horsey-treat behavior, like shoving against you trying to get into a pocket, Cash walks up and smiles.

May I have a treat please?

"Absolutely," say I.

Deep into the first hour of circus-ball training in the round pen, I decided to give us both a short break. I walked across to the far side of the pen and just leaned on the fence, arms draped across the top rail. Cash stood in place, maybe three or four feet away from the ball, and watched me until I was once again looking at him. Then he turned to the ball, took two steps, and nudged it not once, not twice, not three times, but four times, moving it a good ten or twelve feet. Then he turned back to me with that familiar cocked-head question of a look that asked: *Is that worth a paycheck?*

I laughed so hard I could hardly muster a "Good boy."

He got his treat.

And because he was actually having fun, I got mine.

A few days later we were actually playing "pitch and catch." I would roll the ball to him, and he would roll it back. How cool is that?

Thanks, Allen, for continuing our amazing journey of discovery. Our horses owe you.

As do we.

25

New Life

I n the center of the herd the matriarch lay on her side, squeez-
ing new life out of her body. It was a dangerous time for the stallion, and for the herd, for there could be no running if attacked.
New life was more important than old. Perpetuation was everything. Survival of the species.

The stallion could see for miles in every direction. It had been
planned that way. A predator would have to expose itself for quite
some time before it could reach the herd. Not a tactic many
predators cared for.

The foal, once born, would be standing before the end of
its first hour of life, eating and walking by end of hour two, and

running and kicking before hour three came to a close. By hour four, the herd could be once again on the move if necessary. This was encoded for survival over millions of years.

A few days before, another foal had been born, a female, a near copy of her father. Golden, with a big white blaze on her tiny forehead. From under a belly, she was peeking out of the herd, gazing with anticipation at her daddy, wanting to come out and play. But the big stallion pinned his ears, warning her to get back to the center, back to safety. Then he heard the softest of squeaks, and the herd began to stir. He threaded his way through the tightly packed band of horses for his first look at his new son. The foal was sprawled on the ground next to his mother. He had the matriarch's color, a dark bay. No gold at all. And the tiniest white speckle on his forehead. His father's movement drew his attention, and he cocked his head as if he recognized him, and maybe he did. And the stallion was proud.

26

Uh-oh

Hⁿow could this have happened?

We were less than two years into this brand-new world and somehow we had already become one of *them*.

The old school.

The know-it-alls.

Like the farrier with the gutter vocabulary. Or Mariah's cowboy with the long spurs.

"We are not one of *them*," Kathleen argued. "It's not *that* bad."

"Is too," I said. "We should be ashamed of ourselves."

We were sprawled in our favorite chairs after the kids were in bed, reading an article about a safety measure called the Cavalry

Stop. Well, that's not exactly true. We were *re*reading an article about the Cavalry Stop. After having completely dismissed it only a few weeks before.

I've always found it interesting how, at just the right time, God has a way of reaching down and slapping me in the face to remind me of my stated mission.

The serendipity of it all made me smile.

I was fortunate that I could.

Beware what follows.

This is how stagnation begins.

Back in our early horse days, at least a year ago, we were learning an exercise called the one-rein stop. The one-rein stop teaches your horse to stop immediately, right now, no matter what might have just scared the knots out of his tail. Many perfectly calm, happy horses have been known to do all sorts of crazy things when something causes them to leap to the panic side of their brain. Like when Pocket took off with Kathleen because the pit bulls next door came racing down to the fence barking. Kathleen was just learning the one-rein stop, but it wasn't yet automatic. So she just held on, hoping Pocket wouldn't leap over the fence at the far end of the arena.

She didn't.

"Did you try the one-rein stop?" I queried.

She hadn't.

And she didn't appreciate my asking at that moment.

Unfamiliar surroundings, like out on a trail, are prime places for a horse to encounter a perceived predator. Like a plastic bag blowing in the wind. A duck skittering by. More dogs. The logical plan is to be prepared. Just in case. The one-rein stop is designed to keep the rider from winding up as a tree ornament, or worse.

Clinician Clinton Anderson calls it the all-purpose emergency brake.

The horse is taught, over time, to give lightly to one rein, flexing his neck, bending his head all the way back to the saddle, causing his hindquarters to pivot in the opposite direction. The hindquarters are where the horse's power is. This pivoting or disengaging of the hindquarters stops the engine that is propelling him. It takes the steam out of a reactive flight mechanism such as bolting, bucking, or running away. The horse turns in one or two ever-tightening circles until he comes to a stop and can once again engage the thinking side of his brain.

We practiced and practiced one-rein stops in the arena. It's not as easy as it sounds, especially when you're new at everything about horses and are still trying to find your way. Even more especially when you're Kathleen. After her experiences with Mariah and Pocket, she was terrified that some day, somewhere, some horse was going to try to run off with her. This was before Texas, before Skeeter.

I was blowing it because when my horse needed stopping, I was supposed to choke way up on the turning rein, dropping my hand about halfway down the length of the rein, to a point perhaps only two feet away from the horse's mouth. The purpose of this maneuver is to ensure there is enough leverage on the rein in case the horse doesn't respond as lightly as he was taught back when there was no horse-eating squirrel racing across the trail.

Unfortunately when something freaks your horse the last thing you're thinking about is choking up on the rein! The key word there is *thinking*. Generally, when something like that happens, the first reaction isn't to *think*, at all.

The object, then, is to practice this art of choking up over and over until it becomes second nature. Automatic. Pure habit. And since we all hope that there are not that many opportunities to practice when the horse is truly freaked out, I began to use the one-rein stop as a schooling device with Cash. And, of course, the more the one-rein stop was used in other training, the better emergency brake it became. And the more automatic it became for me to choke up on the rein.

I encouraged Kathleen to do the same. To practice, practice, practice. After a time she began to encourage me to shut up. Especially after Pocket had scooted off to the other end of the arena.

"Don't say it," she said.

She wasn't smiling.

The moral of this story, so far, is that we worked long and hard on the one-rein stop, so we could always be safe no matter what. Kathleen practiced even more hours than I did.

We had a big investment in the procedure. And we were finally feeling good about how automatic it had become. We were actually beginning to feel safe. Secure.

Then an issue of *Western Horseman* arrived with a cover story by clinician Curt Pate questioning the safety of the one-rein stop.

I almost fell out of my chair.

When I read the cover headline to Kathleen, she offered to burn the magazine.

One page at a time.

I read the first few paragraphs. His rationale was that pulling a horse into a tight circle, especially when he's going fast, throws the horse off balance, and once he loses his balance, in such a tight circle, he cannot regain it.

Neither of us had ever had any such problems in our arena. As you acquire mileage, you begin to learn how much of what is needed when. How big a circle to turn depending upon how fast the horse is going. I thought, *Of course you're not going to pull him into a tiny, tight circle if he's going really fast.* That would be stupid.

So, unfortunately, I skipped ahead in the article and completely missed the part where Curt says, "Granted, the rider might have developed some sense of security after countless perfectly planned and executed one-rein practice stops in the round pen or arena, but reality outside the pen can be a two-foot-wide mountain trail with a cliff on one side and trees and rocks on the other. In that situation, there's no way to do a one-rein stop without a wreck."

I didn't see that.

I had skipped ahead to scan Curt's alternative, the Cavalry Stop, to see what it was all about. Clinicians are always promoting their clinics, seminars, and presentations, and the Cavalry Stop sounded to me like something designed to give Curt a point of difference for promotion's sake. Something other clinicians didn't have.

I deduced, with no further study or trial, that the Cavalry Stop might not work in an emergency with a rider as nervous as Kathleen at the reins. She needed assurance. And if I had told her, just moments after she had finally become comfortable with her one-rein stop, that she needed to dump it and learn something new, I suspect I would've been looking for a new wife. Or worse.

On to the next article.

We're good with the one-rein stop, thank you very much. No need to learn anything new. Let's just keep those brains closed.

This is where I became acquainted with *higher* learning.

Slap!

The trail we were on was a bit wider than the two feet Curt mentioned in his article. We had at least four or five feet. Thankfully there was no cliff, just a short drop-off on one side, and a hill on the other. Kathleen and I had just finished our picnic lunch, which was shared with our horses, and we were having a perfectly marvelous time.

Well, except for the nuisance of Cash and Skeeter wanting to compete. And stay together. Whenever one of us would canter off, the horse left behind would have a hissy fit to keep up. Or catch up and pass. We were, for the most part, using these opportunities to school on the trail. To remind them that *we* were the leaders, and the desired speed would be determined by us.

I was soaking up the scenery, paying little attention to much of anything, when it happened.

Kathleen decided to canter.

I didn't notice.

Cash did.

Getting no instruction to stay or go, he practically leaped out from under me, rocketing from a slow walk to a fast canter in a split second. I asked for a stop and didn't get it, so instinctively I went for the one-rein stop.

We stopped all right.

The circle I pulled Cash into forced him up into the soft dirt on the hill above the trail. He promptly lost his balance and went down on his side.

Thankfully there were no rocks, just soft dirt and a hill to fall against, so you could say it was only *half* a fall. My instincts must've been in good working order because when Cash plopped into the dirt, I was already out of the saddle, and landed just above him, with only one boot slightly under his body.

He was up in a flash, racing away, back toward our picnic hideaway. Mercifully, the reins were nicely looped over the saddle horn. I have no idea how that stroke of good luck occurred.

Kathleen was about to disappear at a canter around a bend in the trail, oblivious to what had just happened. If she made it around the bend, there was no telling how long it might be before she would look back.

The moment I hit the ground, I yelled as loud as I could, trying to snag her attention before she vanished. It worked. She glanced over her shoulder just in time to catch a glimpse of the situation before she was consumed by a grove of scrub oaks. I quickly climbed to my feet, knowing she was probably having a coronary wondering if I was okay. She reappeared as white as the mane on her palomino.

"I'm all right," I shouted.

She froze in place, mouth agape. The reason was galloping up behind me. Cash had reversed his direction and was now headed toward us at full throttle.

He raced right past me. Eyes wide. Nostrils flared. Definitely way over on the panic side of his brain. He passed Kathleen and Skeeter like a Thoroughbred at the track and disappeared around the bend in the trail. If he made the correct turn, the trail would ultimately lead him back to the horse trailer, which was maybe three miles away. I was checking all my limbs and digits when Kathleen trotted up.

"You want to take Skeeter and go after him?"

I thought for a moment and shook my head.

"No."

It was the right answer, but not necessarily for the right reason.

Skeeter could neither outmaneuver nor outrun Cash.

But what to do? Cash could get lost. Stolen. Hurt himself. If he went far enough, he could be out on public roads. But I knew that chasing him wasn't the right answer. Chasing any animal will only send him away farther, and faster.

The right answer appeared back at the bend in the trail. Cash jolted to a stop and stood there for a very long time, eyes wide, huffing and puffing, but trying to come down off the adrenaline, trying to figure it out.

We are his herd, I reminded myself. *Skeeter, Kathleen, and I are his herd. Stay calm. Let him work it out.*

I asked Kathleen to ride slowly back up the trail, away from Cash, find a nice patch of grass, and let Skeeter munch away. Cash watched. But didn't move. The wheels were turning. He was crawling back to the thinking side of his brain. He glanced back around the bend in the trail. I prayed that no other people were coming along. Finally, he trotted toward me, and for a moment I thought we might re-create Shy Boy's return, but then he broke into a canter and carved a wide semicircle around me, up the hill, then back down again. I noticed that his eyes were no longer saucers. He had a plan.

He slowed to a trot, then a walk as he approached Kathleen and Skeeter, stopping not ten feet away from them. He snorted once and began munching grass.

"What do you want me to do?" Kathleen called.

I considered asking her to reach for the reins, but thought better of it. A miss could mean starting over. And I was now back on the thinking side of *my* brain. And curious.

"Just hang out," I called back. "Take a nap."

Take the time that it takes.

That's a mantra of almost every clinician we've studied. Going slower is faster. Trying to hurry will always take longer.

I didn't move for quite some time.

When I did, it was very casual. And straight toward Skeeter, not Cash, all the while trying to shove my adrenaline through the bottoms of my feet. As I approached, Cash glanced at me but continued to eat. When I reached Skeeter, I rubbed his face and turned my back partially toward Cash, shoulders slumped.

He continued to munch.

I took a couple of steps backward, paused, then another step. My hand stretched out behind me. The hairs of his nose were soon tickling my fingers. I felt a lip nibble, then a breath of warm air.

It's okay, Dad. I'm back now.

I turned and gave him a rub. Then my hand closed around the rein. I cleaned some eye boogers out of his right eye. And it was over.

Choice wins again.

His choice.

Thinking like a horse.

Understanding his fears.

Letting him work it out.

The one-rein stop would continue to have a place with us. But only in the arena. For schooling. It would no longer be our emergency brake. We made a vow out there on the trail to go back home and read Curt Pate's article, from first word to last. At least twice. Maybe more. The article would, as it turned out, make an immense amount of sense. It was very logical. And we would soon begin to work on the Cavalry Stop.

Curt believes that a horse can think of only one thing when confronted by fear. When things go wrong, his primary thought is to find straightness and balance so he can fight or flee. In that respect, bending him in a one-rein stop is an ineffective response to the horse's needs.

"And it doesn't benefit the rider trying to develop enough confidence and balance to ride through problems," Curt says. The Cavalry Stop was developed to teach new cavalry recruits who were inexperienced with horses how to stop in an emergency. Kathleen and I agreed that if it can work for them, it should certainly work for us.

I was ashamed that I had brushed it off so cavalierly. Both Cash and I could've been seriously hurt. Curt speaks in his article about several people who have broken their necks using the one-rein stop in bad situations. We were very fortunate.

The big lesson, however, was all about keeping our minds open and receptive to new and better ideas. When we stop learning, we really stop living, because nothing propels us more effectively through life than knowledge. Knowledge is king. And when we become so stuck in our ways that we ignore available knowledge because we think we don't need it, or we won't go looking to see what's new, or, even worse, refuse to put the knowledge we have to the test of trial, then how are we any better than those who said the earth was flat?

I was not proud of the way I had ignored Curt's article. But the lesson I learned was effective. I would not make the same mistake again. And I was very proud of the way we handled Cash's escape to freedom. It was, in some ways, like sending your child off to college for the first time. You worry about whether all the stuff you've taught him has actually stuck. Did *any* of it stick? Will he make the right choices?

It was a difficult decision to do nothing out there on the trail. But by doing nothing we were, in fact, doing something. We were proving once again, this time very dramatically, that the knowledge and philosophies we had accumulated were working. That we were a good herd, good leaders, and were doing at least some of the right things for the horses.

Cash had told us so.

27

Coming Down

The bay foal was becoming a colt.

And what a fine colt he was. He would prance through the herd, head high, tail arched, like a budding monarch, and the golden stallion was proud of the way he was developing. He had a playful personality and was very confident, which seemed to leave him no need for meanness. The stallion had never seen him bite or kick or be ugly to any of his peers, a trait not often found in feisty young colts. And he had a funny little twist of his head whenever his father would do something the colt didn't understand.

He was very bright, as bright as his father. And the stallion was proud of that as well. He might have wondered if the colt had become friends with a mountain goat, for he was often climbing boulders no one else would venture near, just to retrieve one juicy sprig of dandelion or thistle.

He would walk through a group of his siblings and half siblings and with no more than a look, a twitch of an ear, or the smallest flip of his head get respect and movement from everyone.

Almost everyone.

One sorrel colt, several months older than the bay, always wanted to fight but the bay would have none of it. He just ignored him, and played mostly with his golden half sister. Nothing could frustrate a young fighting stallion like the sorrel more than an opponent who would neither fight nor obey. It amused the older stallion, probably because he had used the tactic himself in his younger days.

But at the moment, the big palomino was pacing and pawing, clearly worried about the bay colt. The young horse was perched atop a high boulder with no apparent way down. The stallion had no idea how the colt had gotten to the top in the first place, likely drawn by some tasty plant. The matriarch, ready to move on, was snorting at him, and pawing, but fear was building and the bay seemed caught up in it, unable to move. Several members of the herd were watching, waiting to see what would happen. The young sorrel couldn't be bothered. He was nipping the last few blades of grass from under a rock.

The stallion eased up next to the matriarch and issued a call of his own to the young bay. Had he misjudged this colt's intelligence?

The matriarch turned away, not pleased with the scents on the afternoon breeze. There was nothing specific, but all was not right and she wanted to move to open spaces. She began to prod the herd toward the east. The stallion lingered, watching the frightened bay. It was important that the colt not get tangled up in reactive activity. He needed to focus on the problem. If he made it down, it would be a good lesson that would serve him well in the future. Finally, the stallion snorted and turned away, leaving the colt alone with his dilemma.

The youngster whinnied and began to blow, his fear building. He knew that he couldn't stay on top of the boulder, easy prey for a wolf or a cat. He needed to be down where he could run. He needed to be with the herd. He needed to figure it out now, and to do that he needed to stop snorting and blowing and start thinking.

He couldn't return the way he had climbed up because the jagged face of the boulder that allowed footholds going up would work against him going down; if he were to fall, they would rip his skin like the teeth of a wolf. So he picked the smooth side of the boulder even though it was steeper. He eased out one foot at a time until he was right at the balance point where one more step would send him to the bottom. He set his rear feet for a jump, squatted just a bit, then leaned forward just enough to start sliding. The next few seconds seemed like forever as he slid down the boulder, squatting even lower as he got closer to the ground. Right where instinct told him, his hind legs pushed against the boulder with all his strength, projecting his body out parallel to the ground. He reached and hit dirt at a dead run.

In a matter of moments he pulled up beside his father, who was trailing the herd toward the east. The great stallion glanced at him and snorted, as if he had expected no less.

28

Mouse

"We don't need another horse!"

The tone was emphatic, finality dripping from every word.

"But sweetie . . . ," I sniveled into the phone, "this one's had so many problems. She *needs* us."

"We already have six horses."

"But we don't have a baby."

"She's a baby???"

There was a long moment of silence.

"Wipe that smile off your face," she said.

Kathleen knows me too well.

Mouse was just under a year old, an abused, neglected creature who had been shipped to Monty Roberts along with three others from the Animal Rescue League of Iowa to benefit from his demonstrations and behavior rehab. My stepson Dylan and I were auditing a weeklong course that Monty was teaching at his Flag Is Up Farms in Solvang, California. That's how we had come to know this bedraggled little filly.

Carol Griglione, president of the rescue league, told me the 4 horses sent to Monty were part of a group of 14 that had been placed in the league's custody by the court, extracted from simply awful living conditions. Mouse, two other yearlings, and a mare who was blind from malnutrition were in a dry lot together starving when the league got to them. None of the horses had food or water. Two of the 14 couldn't be saved. And there were 140 more they expected to get within thirty days. Thank God for people like Carol and her team. One stallion had been found leg deep in his own poop in a tiny little enclosed stall. His name was Defiance. I didn't wonder why.

Little Mouse's feet looked like something out of a horror movie. Way too long in front and turned up in a curl like an elf shoe with cracks and chips galore. She was malnourished, skinny; her mane and tail were tightly matted and filled with burrs and stickers. It had taken six people to herd her into a small corral in Iowa to ready her for the trip to Monty's, and then she jumped the corral! She was frightened to death of humans, until Monty went to work. Using his amazing understanding, and her own language, it took him less than ten minutes to have her following him around the pen; soon he was rubbing her all over, even lifting her feet. It was easy to see that she was hungry for a leader, a compassionate partner. And that deep down she was generous, willing, and very bright. I was in love.

"She needs a good home," I whispered into the phone, not wanting to interrupt Monty's demonstration.

"So do we," Kathleen said, "and we aren't going to be able to pay the mortgage if we keep adding horses."

"I'm sending you a picture I just took with my cell phone."

"I don't want to see it."

But she saw it, and another I sent of a rear foot.

"Oh, the poor baby," she said.

And that was pretty much it. Mouse would be in the third stall of our three-horse trailer when we left on Sunday. Monty had asked me to bring Cash up to Solvang so students could see what an idiot with only two years' experience could do after starting with Join-Up. He didn't actually put it that way, but I'm sure that's what he meant. I resisted, because the last thing I wanted to do was to trot Cash out in front of a bunch of students who had been watching the master at work for several days. But in the end, I relented, if they'd allow me to bring along a buddy to keep Cash company. Kathleen was busy with an upcoming court case (did I mention that she's a lawyer?) and couldn't come along, so Dylan and Handsome filled in for her and Skeeter.

I have wondered whether I would have adopted Mouse had the horse trailer not been with us. Had it not been so easy to get her home. Kathleen says now that she considered telling me to take a third horse so she wouldn't have to worry about something like this happening.

"Well, if you had been there like you were supposed to be . . ."

"It wouldn't have changed a thing." She sighed, a reluctant smile spreading across her face.

I can only conclude that it was meant to be.

As Monty was finishing his first demonstration with Mouse,

the students were all abuzz over the young filly. Several were asking if she was available for adoption. I slipped off the back end of the viewing stand and headed back to the classroom where the rescue league phone number was posted. I wanted to be first on the list. Suddenly there was a hand on my shoulder. One of Monty's instructors.

"Monty would like you and Cash in the round pen."

"Now?"

"That's what he said."

Just great.

My adrenaline was going up. It needed to be down.

And I didn't need to be distracted when I walked in with Cash. He was already wired and nervous with all the unfamiliar horses around. I took several deep breaths.

Monty's round pen is completely enclosed so the horse cannot see out, and no one can see in. The students all stand on an elevated walkway encircling the pen six or seven feet off the ground. When the big wooden door swung open, Cash's eyes looked like saucers. So did mine, I'm sure. There were fifty or sixty pair of eyes scattered around the pen. All fixed on the two of us. And one pair belonged to Monty Roberts, a man who was much better at what I was doing than I would ever be. I've spoken to hundreds, even thousands, of people before, but on this day I was way out of sync.

Thank goodness for Monsieur le Cash. And Join-Up. And all the learning we had done together. Because in the end Cash did everything he was supposed to do as if it was second nature, totally ignoring my raised adrenaline level. He moved his hindquarters this way and that with no more than a glance from me. And his forequarters went left and right with a wiggle of a finger. He backed up and came forward, sidestepped both ways, and lunged (trotted) at the end of the rope, changing directions with a mere point of my finger. And, on cue, he gave everybody a big fat smile. A lopsided smile that turned up on one side. Who says Elvis has left the building?

Near the end of the session, I unhooked the lead rope and turned Cash loose. He trotted straight toward Monty and stretched up for a rub. He had never met Monty before, but somehow he knew intuitively which one was *the man*.

I spent much of the time in the pen just standing, talking to the students, answering questions, and Cash was always right at my shoulder, blowing in my ear and checking out my nose. And that's what the class appreciated the most. Not the various maneuvers he had learned to do flawlessly, but rather that the two of us clearly had a very special relationship. Kathleen suspects that's exactly what Monty wanted them to see: the kind of relationship that is born in Join-Up.

"It was so obvious," many of them said afterward. "This horse loves you."

I thought to myself how easy it is to get so wrapped up in the task at hand, the bit of training, the trick, the discipline, that we forget about the most important part of the relationship with our horse, or anyone for that matter. The relationship itself. When you get that right, the rest is easy. I turned to Cash and rubbed him on the forehead.

"What a good teacher you are," I said.

I slipped away from the group of students who were admiring my big handsome partner.

"Excuse me." I said. "I need to make a phone call."

Just a few weeks before I had been talking to Kathleen about how our leap into horses had been so feverish and obsessive that I was finding it difficult to remember those early days and the hours I had spent with each of our horses in the round pen and the arena. The relationship building, the training, the learning by all of us. All crammed and compressed into every waking moment. And all seemingly so automatic now. So accepted. So taken for granted. I thought it would be a good exercise to start anew with a horse we didn't know, now that we had so much more understanding.

"Almost two whole years," Kathleen chirped.

"You know what I mean," I said.

"Not exactly," she said. "What *do* you mean?"

"I suppose I must need some sort of confirmation that we don't have six flukes."

"You mean we just got lucky?"

"I suppose."

"Six times?"

"It'd be nice to know."

"We know," she said. "We don't need another horse."

It wouldn't be long before I would hear that line again.

After putting Cash back in the turnout with Handsome, I went straight to the classroom and called Carol Griglione at the Animal Rescue League. During that conversation Kathleen and I received yet another chance to prove the concepts of relationship from the horse's end of the lead rope—an undernourished, bad-footed, scraggly, sweet, bright, cute, fantastic little filly who had never had as much as a nice hello from a human prior to meeting Monty Roberts.

Kathleen was right. We had no business adopting her. Who knew what she had been exposed to in the awful conditions where she had been living? And it was difficult enough to keep our other six moving forward, never mind, as Kathleen had said, the cost of the feed and the added poop to muck. And up to this point, the youngest horse we had any experience with was six years old.

Sojourn.

Who was no longer with us.

And he had spent several months with a trainer before arriving at our place.

Mouse was barely a year old and had had no training whatsoever, except the time she spent with Monty. Maybe three hours total over the week.

Still, I was smitten.

But the minute I faxed the adoption agreement back to Carol, adopter's remorse hit me like a brick in the back of the head. What did I know about baby horses? What if it didn't work out? Already everyone in the course was asking how soon updates on Mouse would be on our website. Suddenly I was very nervous. Maybe this was all wrong.

Maybe I should take Mouse out for a walk and see how she responds to someone other than Monty.

Dylan and I worked with her in her pen, generating a bit of *followup*, the next step after Join-Up. Encouraging the young horse to follow us around, which she did very readily. Then we walked her down to meet Cash and Handsome, and that went well. Lots of nose-to-nose sniffing and blowing. So I was beginning to feel better about the whole thing.

Until feeding time.

Mouse was chomping away in her pen. I went in to give her a rub and say good night, but suddenly met her hindquarters spinning toward me and a pair of striking feet! I twisted my shoulder leaping toward the gate, barely avoiding a hoof to the thigh. My adrenaline soared and I was shaking. This was a different horse! Once outside, I was caroming wildly from fear to anger to not really knowing what to think.

But there was no question I had made a huge mistake.

Was she some sort of Jekyll and Hyde?

Get back in the herd, I kept telling myself. *Why would she flip out like that?*

I dug my way back to the thinking side of my brain.

And finally it struck me. Just look at her. Ribs showing. Spindly legs. Before rescue, there was no telling when she'd last had a decent meal. And maybe she had to deal with bigger, older horses moving her out and taking her food away from her. Or perhaps she had never dealt with a herd at all and had no idea about hierarchy. In any case, she was clearly saying, *This is mine!* Perhaps the best meal she had ever had. And she wasn't giving it up to anyone. *Herd member or not, stand clear when I'm eating!*

Proof once again that we never stop learning.

As I write this, Mouse has already come to realize that there will always be enough food with us. I wondered how many of the people we had met over our short two years with horses would have whipped her for striking at them. Or walked away and left her, with no attempt to understand why she had acted the way she did.

Two days later, back at home, while our natural-hoof specialist was rasping away on one of those same back feet, I was rubbing her, one hand on her forehead and one under her jaw, and she fell asleep in my arms, trusting me to support the weight of her head, confident that no one was going to hurt her.

And I was smiling.

She's going to be fantastic.

It suddenly occurred to me that Mouse was a bridge from one journey of discovery to another. Yet another way to help horses. And I thanked God for that, and promised I would be telling folks about Miss Mouse.

29

Empty Stalls . . . Again

It was late afternoon and Kathleen and I were once again sitting on our front porch looking out over the cute white stalls with the red roofs. The sun was sinking beneath the ridge of mountains to the west. The larger stall had been split into two, so now there was a total of three.

"Those stalls surely seem empty," Kathleen said. "Wouldn't it be nice if there were a couple of horses ambling back and forth down there?"

"Those were the days," I said.

"Like a picture postcard."

We both smiled.

"Too bad it didn't work out. I enjoyed watching them."

"Yeah," Kathleen said. "Too bad."

"It's hard to believe we were so stupid."

"Not stupid," she corrected. "Lacking knowledge."

"That was a long time ago."

"Coming up on two years." She chuckled, raised her wineglass to mine. They clinked softly.

"We done good," she said. "We done good."

Our first three horses occupied those cute little stalls for several months. They all wore metal shoes, ate pellets from a table-high bucket, and hay from a feeder even farther off the ground. They stood in one spot to eat with no reason to move around. In fact, they stood in one spot most of the day. They had nowhere to go. They *did* have time with the herd. Sort of. The three stalls were open to each other and the horse in the middle could actually nip or be nipped by the two who flanked him. Not because we set it up that way. That's how it was when we bought the house. Fortunately, there was no barn, or I'm sure they would've been in it. But there were also no pastures or turnouts, and the only exercise the trio got was when we took them to our small arena. We bought leg wraps. Almost bought blankets. In short, we were the typical horse owners.

Almost.

Stumbling across that article on Monty Roberts saved us. By giving our horses the choice to accept us as their herd member and leader, we created a true relationship with each of our horses. We saw them and talked with them and rubbed them often. We hung out with them, even when there was no riding planned. No agenda. We watched them interact with each other, and learned more and more of their language. And with all of this came responsibility. Lots of it. We *wanted* them to have a good, happy, and healthy life. We did *not* want to merely use them. And we felt responsible for making it so. So we began our quest for knowledge, realizing from the get-go that there were many schools of thought on how to bring up horses. We read books and watched DVDs from Monty, Clinton Anderson, and the Parellis. And we subscribed to horse magazines.

One of the first to arrive in the mail contained the article on barefoot versus shoes. I was dumbfounded. It had never once occurred to me that a horse was designed to stand on his own four feet, without man-made shoes.

More reading.

I gobbled up the works of natural-hoof-care practitioners Pete Ramey and Jaime Jackson. Together they have libraries full of research on wild horse hooves and decades of first hand experience at trimming barefoot horses with the wild horse trim.

Now all seven of our horses (yes, with Miss Mouse, we are seven! Egad!) are barefoot with the wild horse trim and they're doing just terrific. In the arena, on the trail, on the road, wherever.

Some believe that high-performance competition requires shoes to protect the horse's feet from stresses they wouldn't be encountering in the wild. Never mind that if the organizers of high-performance competition actually believe that, it begs the question of why they are doing it to the horse in the first place. But even more important, the barefoot folks have disproved the concept that horses need shoes to rein, or jump, or race, or do any other competition.

But will my horse still win without shoes? After all, winning is why we're in this thing in the first place.

The answer is yes. Every issue of *The Horse's Hoof Magazine* reports on barefoot winners in virtually every discipline. The most recent issue reports on High Flying Princess, a barefoot barrel racer with six first-place wins in her first barefoot season. Houston trimmer Eddie Drabek has many similar stories. As do Pete Ramey and so many others.

But jumpers need cleats on their shoes to get traction for a jump.

The response is no they don't. The Fall 2007 issue of *The Horse's Hoof* features a lengthy article on the Horses First Racing Club in the United Kingdom. All of their hunter/jumpers compete barefoot.

But my horse needs shoes to do a good skid stop because shoes are slick and they slide better.

Ah, so it's okay to restrict the circulation in the rear hooves in order for Flicka to slide farther?

Hmm . . .

The logic and bank of knowledge on this subject were enough for us. We pulled every shoe. But didn't stop there. We kept digging ever deeper to learn what movement around the clock can mean to the horse. What his natural systems, when working on their own without human intervention, can mean to the horse. As well as the gain when he always eats from ground level.

And so it went.

Why are we so different?

We're not, really. There are a lot of folks around the world who believe as we do and are doing the same things we're doing. Perhaps more than you think, but we are still a decided minority. Is it because the majority doesn't really care for their horses? I don't think so. We cared for our horses from the first moment we chose them. But we didn't know. And because we were listening to folks who were just like us, who also didn't know, we were all doing the wrong things. It was when the horses were allowed to choose *us* that things really began to escalate beyond just "caring." That's when a feeling of responsibility really took hold. I realized for the first time that this horse had just said to me, *I trust you,* and I promised him, *I will do right by you. Your needs will take precedence over my desires.*

Dr. Matt was right when he said that until recently horses were pretty much just beasts of burden. But I believe many of them still are. Just different burdens. Instead of pulling a plow or a wagon, they're jumping fences. Or racing. Or doing heavy-handed dressage ballets. That's not to say that those owners don't care for their horses. I suspect they do. Some of them. But Monty Roberts' website details the story of what happened when one of the world's top dressage trainers Joined-Up with her horse. She totally abandoned the heavy-handed approaches that many if not most dressage trainers use and now her horses are performing as a matter of choice. And remember Horses First, that winning race and hunt club in the UK. Or the top reining and cutting trainers in the United States who have taken their horses barefoot, like Clinton Anderson. It's amazing what you can do for your horse when you care enough for him to spend the time to gain the knowledge that will allow him to live longer, healthier, and happier.

It's important for this word to spread. During the past five years the horse population in the United States has leaped upward to ten million, an increase of approximately three million in just a few short years. Overpopulation has flooded the market, and many of these horses are ill cared for. Recent projections indicate that as smaller generations follow the baby boomers, fewer horses are going to be sold. *Horse & Rider* magazine projects that high-end breeding and sales of glitzy foals will fall off substantially and the only market to sustain will be for the family horse. The baby boomers themselves, those who have horses, will want to involve their grandkids—and for that they'll need calm, bomb-proof, babysitting kinds of horses, the kind of horses who have made the choice to bring humans into their herd and into leadership roles.

The old cowboy on the Texas trail ride told Kathleen that all her horse wanted to do was get back to the herd. Kathleen showed him that she *was* the herd. The horse was already there. You don't really need to be a horse to be part of the herd. You just need to spend the time and effort to think like one.

And you need to care.

30

Part of the Herd

The man had been there for days. Sitting on a far-off boulder. Alone. Just watching. He was same man they had encountered in the box canyon, and certainly unlike any man the stallion or the matriarch had ever encountered before. He had no means of transportation other than his feet. He did nothing that seemed to be a threat. He was just there.

A couple of the younger horses had ventured closer, curious about the man. But the stallion had called them back. His bay foal, who was now nearing a year old, was one of the curious. The stallion took such pride in how bright this young colt was, yet one day he would have to shun him from the herd. There could only be one stallion. There are exceptions, but they are rare. When shunned, a young stallion will usually band together with other like youngsters until he is mature enough and strong enough to challenge some older stallion for a herd. But all of this was at least a year away for the young bay colt, maybe more.

Now there was only playful curiosity about the strange man. One cold morning the man moved closer, walking first this way, then that, never straight toward the herd. He found another boulder, much closer, and there he sat. Again, just watching with kind eyes. And for a long while the herd watched him. Especially the stallion. And inexplicably he felt no fear. He continued to graze.

At night the man would bundle up in his blanket and scrunch under an outcropping of the big boulder. And the horses would draw closer, meandering, as if only to reach for one more sprig of grass. One step closer. The man would never look at them. Sometimes during the day he would turn his back and lower his shoulders and head in a show of friendly submission, saying, *I am approachable, I am not a predator.*

The stallion was drawn to the kindness in the man's eyes, and something somewhere deep inside moved him to like this man. But he never ventured close because his job was to protect the herd.

One cold morning, huddled in his blanket under the boulder, the man awoke to the warm breath of horse. He slowly opened his eyes to see the nostrils of the bay colt, sniffing, puffing. He did not look the young horse in the eye, as a predator would, but focused on his nose, and he, too, puffed and sniffed a greeting. The colt stepped back, and the man sat up, wrapping the blanket around his shoulders. He turned his back to the young horse, only partially, his head down, his shoulders slumped, and waited for the colt to touch him again before he reached out and rubbed the young horse on the nose, then on the forehead. He was impressed with the colt's manner, the clarity of his eye, the obvious intelligence.

It would be several more days before he would ask the colt to come with him. Neither the stallion nor the matriarch minded. This man seemed to be one of them.

Before leaving with the colt, the man managed to persuade the stallion to touch his shoulder and accept a rub on the forehead. He then turned and walked away. The bay colt followed. It was days later, back at his ranch, when he chose a name for the colt.

It felt right to call him . . . *Cash*.

31

Synthesis

Discovering the mysteries of the horse is a never-ending journey, but the rewards are an elixir. The soul prospers from sharing, caring, relating, and fulfilling. Nothing can make you feel better than doing something good for another being. Not cars. Not houses. Not face-lifts. Not blue ribbons or trophies. And there is nothing more important in life than love. Not money. Not status. Not winning.

Try it and you will understand what I mean. Apply it to your horses and your life. It is the synthesis of this book and why it came into being.

Give the choice of *choice*. To your horse, or your employee, or your friend, or your loved ones. Care enough to want them to be healthy and happy. It will come back a hundredfold.

And always question everything. Be your own expert. Gather information and make decisions based upon knowledge and wisdom, not hearsay. Know that if something doesn't seem logical, it probably isn't. If it doesn't make sense, it's probably not right.

Learn the art of discipline with compassion.

And care about the way we care for the domestic horse. It needs to change. An extreme makeover, if you will. Going back to square one and beginning anew.

There are many who teach relationship, riding, and training with principles of natural horsemanship. Others support the benefits of going barefoot with the wild horse trim. Still others write that your horse should eat from the ground and live without clothes and coverings. Some promote day-and-night turnout, where your horses can move around continuously. But few have explored how dramatically one without the other can affect the horse and his well-being. Few have put it all together into a single philosophy, a unified voice, a complete lifestyle change for the domesticated horse. When I gave Cash the choice of choice and he chose me, he left me with no alternative. No longer could it be what I wanted, but rather what he needed. What fifty-five million years of genetics demanded for his long, healthy, and happy life.

I'm still astonished when I think of where Kathleen and I began such a short time ago, and where most horse owners still are today, training with dominance and cruelty, cooping up their horses in small spaces, weakening their natural immune systems, feeding them unnaturally, creating unhealthy hooves and bodies with metal shoes. All because most folks actually believe it's the right thing to do.

Yes, there are those who still only want a beast of burden. *Do as I say. Make me a winner. Jump higher. Run faster. Slide farther.* People who care not about having a relationship with their horse, and who will, when confronted, continue not to care about the health and happiness of their horse. But I believe that most horse owners today care about their horses and are operating, as we once were, with little more than emotional logic, old wives' tales, and very little real knowledge. I hope this book will be a crack in the armor, a small breeze if not the strong winds of change, a resource for what needs to be done.

And a longer, happier, healthier life for all horses.

Afterwhinny

Joe has known me since I was nine. I'm almost eleven now. He really doesn't know from whence I came, nor does my former owner. Only that my name, Cash, came with me. Joe believes that the story he told of wild horses throughout this book is a fable. I've heard him say that there's no validating written record. And he's not old enough to have been there for most of it. But I've also heard that very few good stories are pure fiction. There's a place where a storyteller's life, his cares and concerns, his passions, and his imagination all come together into something magical that's part truth, part *could be*, and part *maybe not*. I know Joe, and I know this to be true: His story of my ancestors came from the heart. And from good research. Because any or all of it *could've* happened just as you read it. That's the way it is in the wild. Joe was trying to illustrate how we are supposed to live, and how truly easy it is to *be* one of us, and to allow us to live as we should, as we always have.

I'm sorry that Joe believes his story to be a fable. Especially the part about where and to whom I was born. Because Joe doesn't know. He wasn't there.

And I was.

Acknowledgments

If the last Benji movie, *Benji Off the Leash*, had been a big success, we would've never owned horses and this book would've never been written. The movie was not a big success. It was unable to compete as an independent film against the huge promotional dollars being spent by the Hollywood studios these days. That experience left a huge, gaping hole in my life. I was convinced that *Benji Off the Leash* was going to raise the bar for family films. Be an example that would show Hollywood the error of its ways. It had a strong story that set a good example, without the use of four-letter words, sexual innuendos, or violence. I was certain that God was using *Benji Off the Leash* to prove once and for all that good stories do not need to lower the bar to entertain. It was clear, at least to me, that God had been involved in the movie from the beginning, that He wanted it to be made. The money was raised in record time. We were forced to accept Utah as a production location, against our wishes, but once we were there, many of the usual production problems miraculously vanished. And we found Tony DiLorenzo, a young composer searching for his first movie. He wrote an amazing score that we could never have afforded with a seasoned composer, and I believe Tony will become one of the finest film composers in the business.

Yet with all of that, the film did not do well.

And there was this huge hole to fill.

When depression tries to claw its ugly self into your being, there are but two choices. Give in to it or grab it by its scrawny neck, sling it to the ground, and pull yourself out of that hole.

Growth always seems to arise out of adversity.

I, of course, didn't know it when it was happening, but God was telling me it was time to move on. To fill another need. To make a difference.

If the movie had been even marginally successful, He knew I'd be off working on another one.

But I wasn't to go there.

Instead, I tried to forget by turning to horses. And learning about them. I wanted to fill the emptiness of that dark hole. Then, slowly, Kathleen and I began to realize that something was amiss with the traditional methods of caring for these beasts. And, quite unexpectedly, an amazing journey of discovery lay before us.

A new passion was born.

My first acknowledgment, therefore, is to God for never failing to do whatever it takes to make me listen, no matter how hard I try not to. For the tough love I so often need. For caring that much. And for using me as a humbled instrument of His will.

Next, from the bottom of my heart I thank the investors in

Benji Off the Leash, dear friends all. At best it will be a long time until you recoup your investment, yet I have never lost your support, or your friendship. In addition to funding a terrific movie with a wonderful message, you have inadvertently made a difference for horses everywhere.

During the promotion of the film, one of the publicists set up a radio interview by telephone with Dr. Marty Becker, well-known author, syndicated columnist, radio host, and *Good Morning America*'s vet in residence. A week or so after the interview, Marty called and asked if it would be possible for me to bring Benji to a fund-raiser in his hometown of Bonners Ferry, Idaho. We were in the middle of a coast-to-coast, major-market promotional tour for the film and he was asking that we pause for two days and come to a town of 2,700 people for a benefit screening. "We could also do one in the neighboring town of Sandpoint," he added. A much bigger town, almost 8,000 people.

It was clearly another God thing, because I took one look at Marty and Teresa's beautiful ranch—and their horses—and convinced everyone involved that it would be a nice breather between Seattle and Chicago. Kathleen met me in Spokane and we drove up to Bonners Ferry for a perfectly wonderful two days nurturing a pair of new lifelong friendships.

Why does any of that matter? Because if it hadn't been for the movie, the investors, and God, this most unusual meeting with Marty Becker would have never happened. And if the meeting had never happened, Marty Becker would never have become such a giving and loving friend, and he would've never have introduced me to his literary agent, David Vigliano, easily one of the best in all the world. If I had never met David, it stands to reason that he never would have become my agent and I would certainly be, by this time in the process, completely insane. And, without David, I'm sure the book would've never made it to Shaye Areheart, the most loving publisher on the planet.

So from the bottom of my heart, thank you so much, Dr. Marty Becker. Maybe our horses will be pasture mates yet. And thank you, David Vigliano, for loving the book right from the get-go. For believing. And for saying exactly the right thing every time I needed it. Thank you, Shaye Areheart, for having the faith to put the power of America's largest publishing house behind these meager words of mine. I'm still aghast.

Next has to be my editor, Peter Guzzardi. What a thankless job, trying to pull the best out of a paranoid independent writer and filmmaker who, having always been independent, never once had the blessing of a sensibility like Peter's to make something better. And when there's no one there to try, one can become very possessive, fearful, mistrustful, unreasonable, and obsessed. Yet somehow Peter managed to carefully weave his way through my insecurities and help me make this book so much better. Thank you, Peter. I hope it was somehow not as bad an experience as I imagine. I would hate to have to deal with me.

Thank you, Monty Roberts, for your friendship and for being there with Join-Up as we began this process. If we hadn't given our horses that choice to be with us, right in the beginning, our entire experience would have been sadly different, for it was that moment of Join-Up with Cash that caused me to change from owner to partner. From *like* to *love*. From the boss to a member of the herd and a true leader. You have blessed me with the soul of a horse.

Thank you to all the clinicians, trimmers, vets, and authors listed in the Resources section of this book, many of whom have become friends since those early days not really so long ago.

Thank you for sharing your decades and decades of rich experience that allowed us to get so quickly up to speed, to understand the truth, and to become yet another messenger to carry your mission forward.

Thank you, Cash. Thank you to all of our horses. Each of you has such a wonderfully unique personality, and you have brought so much into our lives. But especially Cash. I wonder if you understand how very much you mean to me. When you cock your head and peer straight into my soul, I believe, somehow, that you do.

Lastly, there is my editor before Peter, the love of my life and my second soul mate, Kathleen. How does one become so fortunate as to have two such intelligent, caring, compassionate soul mates in one lifetime? Whenever I'm buried in a project, Kathleen is always there. If I'm editing a film, she consults every evening on what we've edited that day. And she's so brutally honest that afterward we might not speak for hours. It's difficult, emotional work that deserves combat pay. The same is true with this book. She read every word, every chapter, over and over again.

"I don't see any changes," she might say.

"What do you mean? The third word in the eighteenth paragraph is changed."

Combat pay, indeed.

Thank you so much, Sweetie, not only for your help and support, your love, your glorious ideas, and, yes, for the title of this book, but also for allowing me to tell your side of this journey as it actually was: fearful, frustrating, and embarrassing. I'm sure there were times when you simply wanted to quit and walk away. It is your book as much as mine. I love you so much.

Photograph Credits

Chapter 1
Stallion in the Wild—American West
Photo by Pete and Ivy Ramey
www.hoofrehab.com

Chapter 2
Joe and Cash
Photo by Kathleen Camp
www.thesoulofahorse.com

Chapter 3
Horses in the Wild—American West
Photo by Pete and Ivy Ramey

Chapter 4
Joe and Kathleen's Tack Room
Photo by Joe Camp

Chapter 5
Herd at Ortega Mountain Ranch
Photo by Laurra Maddock
www.ortegamountainranch.com

Chapter 6
Joe and Cash
Photo by Kathleen Camp

Chapter 7
Babies Sleeping—Carpe Diem Farm
Photo by Joe Camp
www.carpediemfarm.com

Chapter 8
Joe and Kathleen's Natural Pasture
Photo by Joe Camp

Chapter 9
A Family in the Wild—American West
Photo by Pete and Ivy Ramey

Chapter 10
Pocket and Sojourn Communicating
Photo by Joe Camp

Chapter 11
Two Wild Horses Communicating—
American West
Photo by Pete and Ivy Ramey

Chapter 12
Cash, Joe, and a Pawleys Island
Hammock
Photo by Kathleen Camp

Chapter 13
Rusting Shoes and Nails from Cash,
Handsome, and Pocket
Photo by Joe Camp

Chapter 14
Dancing at Sunset—Arrowhead
Mountains
Photo by Ginger Kathrens
www.TheCloudFoundation.org

Chapter 15
Pocket—Before We Knew
Photo by Joe Camp

Chapter 16
Stallion in the Wild—American West
Photo by Pete and Ivy Ramey

Chapter 17
Sophie Having a Blast—Ortega
Mountain Ranch
Photo by Laurra Maddock

Chapter 18
Joe and Cash
Photo by Kathleen Camp

Chapter 19
Family Band in Snowstorm—
Arrowhead Mountains
Photo by Ginger Kathrens

Chapter 20
Kathleen and Skeeter
Photo by Joe Camp

Chapter 21
Wild Herd—American West
Photo by Pete and Ivy Ramey

Chapter 22
Joe and Kathleen's Remaining
Hitching Post
Photo by Joe Camp

Chapter 23
Sojourn on the Run
Photo by Joe Camp

Chapter 24
Circus Ball Going for a Ride
Photo by Joe Camp

Chapter 25
Going to the Water Hole
Photo by Ginger Kathrens

Chapter 26
Joe Flexing Cash
Photo by Kathleen Camp

Chapter 27
Members of a Wild Herd—American
West
Photo by Pete and Ivy Ramey

Chapter 28
Miss Mouse—Shortly After Coming
Home
Photo by Joe Camp

Chapter 29
An Empty Stall—Handsome's Lead
Rope
Photo by Joe Camp

Chapter 30
Joe with Skeeter, Pocket, and Cash
Photo by Kathleen Camp

Chapter 31
Joe with Cash
Photo by Kathleen Camp

Resources

There are, I'm certain, many programs and people who subscribe to these philosophies and are very good at what they do but are not on the following list. That's because we haven't experienced them yet, and we will only recommend to you programs that we believe, from our own personal experience, to be good for the horse and well worth the time and money.

NATURAL HORSEMANSHIP

This is the current buzzword for training horses or teaching humans the training of horses without any use of fear, cruelty, threats, aggression, or pain. The philosophy is growing like wildfire, and why shouldn't it? If you can accomplish everything you could ever hope for with your horse and still have a terrific relationship with him or her, and be respected as a leader, not feared as a dominant predator, why wouldn't you? As with any broadly based general philosophy, there are many differing schools of thought on what is important and what isn't, what works well and what doesn't. Which of these works best for you, I believe, depends a great deal on how you learn, and how much reinforcement and structure you need. We have more or less shuffled together the first three whose websites are listed below, favoring one source for this and another for that. Often, this gives us an opportunity to see how different programs handle the same topic, which enriches insight. But, ultimately, they all end up at the same place: When you have a good relationship with your horse that began with choice, when you are respected as your horse's leader, when you truly care for your horse, then, before too long, you will be able to figure out for yourself the best communication to evoke any particular objective. These programs, as written, or taped on DVD, merely give you a structured format to follow that will take you to that goal.

www.montyroberts.com Start here, please. Learn Monty's Join-Up method, either from his books or DVDs, on sale at this website address. Watching his *Join-Up* DVD was probably our single most pivotal experience. Even if you've owned your horse forever, go back to the beginning and watch this DVD, then do it yourself with your horse or horses. You'll find that when you unconditionally offer choice to your horse and he chooses you, everything changes. You become a member of the herd, and your horse's leader, and with that goes responsibility on his part as well as yours. Even if you don't own horses, it is absolutely fascinating to watch Monty put a saddle and a rider on a completely unbroken horse in less than thirty minutes (unedited!). We've also watched and used Monty's *Dually Training Halter* DVD and his *Load-Up* trailering DVD. And we loved his books: *The Man Who Listens to Horses, The Horses in My Life, From My Hands to Yours: Lessons from a Lifetime of Training Championship Horses*, and *Shy Boy: The Horse That Came in from the Wild*. Monty is a very impressive man who cares a great deal for horses.

www.parelli.com Pat and Linda Parelli have turned their teaching methods into a fully accredited college curriculum. We have four of their home DVD courses: Level 1, Level 2, Level 3, and Liberty & Horse Behavior. We recommend them all, but especially the first three. Often, they do run on, dragging out points much longer than perhaps necessary, but we've found, particularly in the early days, that knowledge gained through such saturation always bubbles up to present itself at the most opportune moments. In other words, it's good. Soak it up. It'll pay dividends later. Linda is a good instructor, especially in the first three programs, and Pat is one of the most amazing horsemen I've ever seen. His antics are inspirational for me. Not that I will ever duplicate any of them, but knowing that it's possible is very affirming. And watching him with a newborn foal is just fantastic. The difficulty for us with *Liberty & Horse Behavior* (besides its price) is on disk 5, whereon Linda consumes almost three hours loading an inconsistent horse into a trailer. Her belief is that the horse should *not* be *made* to do anything; he should *discover* it on his own. I believe there's another option. As Monty Roberts teaches, there is a big difference between *making* a horse do something and *leading* him through it, showing him that it's okay, that his trust in you is valid. Once you have joined up with him, and he trusts you, he is willing to take chances for you because of that trust, so long as you don't abuse the trust. On his trailer-loading DVD, Monty takes about one-tenth the time, and the horse (who was impossible to load before Monty) winds up loading himself from thirty feet away, happily, even playfully. And his trust in Monty has progressed as well, because he reached beyond his comfort zone and learned it was okay. His trust was confirmed. One thing the Parelli program stresses, in a way, is a follow-up to Monty Roberts's Join-Up: You should spend a lot of time just hanging out with your horse. In the stall, in the pasture, wherever. Quality time, so to speak. No agenda, just hanging out. Very much a relationship enhancer. And don't ever stomp straight over to your horse and slap on a halter. Wait. Let your horse come to you. It's that choice thing again, and Monty or Pat and Linda Parelli can teach you how it works.

www.downunderhorsemanship.com This is Clinton Anderson's site. Whereas the Parellis are very philosophically oriented, Clinton gets down to business, with lots of detail and repetition. What exactly do I do to get my horse to back up? From the ground and from the saddle, he shows you precisely, over and over again. And when you're in the arena or the round pen and forget whether he used his left hand or right hand, or whether his finger was pointing up or down, it's very easy to go straightaway to the answer on his DVDs. His programs are very task-oriented, and, again, there are a bunch of them. We have consumed his *Gaining Respect & Control on the Ground, Series I* through *III* and *Riding with Confidence, Series I* through *III*. All are multiple DVD sets, so there has been a lot of viewing and reviewing. For the most part, his tasks and the Parellis' are much the same, though usually approached very differently. Both have served a purpose for us. We also loved his *No Worries Tying* DVD for use with his Australian Tie Ring, which truly eliminates pull-back problems in minutes! And on this one he demonstrates terrific desensitizing techniques. Clinton is the only two-time winner of the Road to the Horse competition, in which three top natural-horsemanship clinicians are given unbroken horses and a mere three hours to be riding and performing specified tasks. Those DVDs are terrific! And Clinton's Australian accent is also fun to listen to . . . mate.

THE THREE PROGRAMS above have built our natural-horsemanship foundation, and we are in their debt. The following are a few others you should probably check out, each featuring a highly respected clinician, and all well known for their care and concern for horses.

www.imagineahorse.com This is Allen Pogue and Suzanne De Laurentis's site. Allen's work has unfortunately cast him as a trick trainer, but it's so much more than that. We've just recently discovered Allen and are dumbfounded by how his horses treat him and try for him. His work with young horses is so logical and powerful that you should study it even if you never intend to own a horse. Allen says, "With my young horses, by the time they are three years old they are so mentally mature that saddling and a short ride is absolutely undramatic." He has taken Dr. Robert M. Miller's book *Imprint Training of the Newborn Foal* to a new and exponential level.

www.johnlyons.com John Lyons's work is terrific, and he is very well respected. But his system is entirely different from the preceding three (which demonstrates that there are many ways to communicate with horses, and once you have a good foundation, you can pretty much develop the type of communication best suited to you and your horse). To follow John's system, you have to begin at his beginning and stick with it. If you try dropping into the middle, it's like trying to understand Greek.

http://users.elknet.nct/circlewind/buster.htm We stumbled onto Buster McLaury at the Texas trail ride mentioned earlier. Very much a natural horseman, he gave a wonderful demonstration. Buster's website chronicles his writings in various horse publications and contains his clinic schedule. He apparently has no DVD programs at the moment, but if he's in your area, we recommend a look-see.

www.robertmmiller.com Dr. Robert M. Miller is an equine veterinarian and a world-renowned speaker and author on horse behavior and natural horsemanship. I think his name comes up more often in these circles than anyone else's. His first book, *Imprint Training of the Newborn Foal*, is now a bible of the horse world. He's not really a trainer, per se, but a phenomenal resource on horse behavior. He will show you the route to "the bond." You must visit his website.

TAKING YOUR HORSE BAREFOOT

Taking your horses barefoot involves more than just pulling shoes. The new breed of natural-hoof-care practitioners have studied and rely completely on what they call the wild horse trim, which replicates the trim that horses give to themselves in the wild through natural wear. The more the domesticated horse is out and about, moving constantly, the less trimming he or she will need. The more stall-bound the horse, the more trimming will be needed in order to keep the hooves healthy and in shape. Every horse is a candidate to live as nature intended. The object is to maintain their hooves as if they were in the wild, and that requires some study. Not a lot, but definitely some. I now consider myself capable of keeping my horses' hooves in shape. I don't do their regular trim, but I do perform interim touch-ups. The myth that domesticated horses *must* wear shoes has been proven to be pure hogwash. The fact that shoes degenerate the health of the hoof and the entire horse has not only been proven but is also recognized by even those who nail shoes on horses. Successful high-performance barefootedness with the wild horse trim can be accomplished for virtually every horse on the planet, and the process has even been proven to be a healing procedure for horses with laminitis and founder. On this subject, I beg you not to wait. Dive into the material below and give your horse a longer, healthier, happier life.

www.hoofrehab.com This is Pete Ramey's website. If you read only one book on this entire subject, read Pete's *Making Natural Hoof Care Work for You.* Or better yet, get his new DVD series, which is fourteen-plus hours of terrific research, trimming, and information. He is my hero! He has had so much experience with making horses better. He cares so much about every horse that he helps. And all of this comes out in his writing and DVD series. If you've ever doubted the fact that horses do not need metal shoes and are in fact better off without them, please go to Pete's website. He will convince you otherwise. Then use his teachings to guide your horses' venture into barefootedness. He is never afraid or embarrassed to change his opinion on something as he learns more from his experiences. Marci Lambert, our natural trimmer, and Pete are very much in sync, and our horses are all barefoot and all terrific. Pete's writings have also appeared in *Horse & Rider* and are on his website, along with his clinic schedule, which takes him all over the United States and Europe. Recently he has taken all of Clinton Anderson's horses barefoot.

THE FOLLOWING ARE other websites that contain good information regarding the barefoot subject.

www.TheHorsesHoof.com This website and magazine of Yvonne and James Welz is devoted entirely to barefoot horses around the world and is surely the single largest resource for owners, trimmers, case histories, and virtually everything you would ever want to know about barefoot horses. With years and years of barefoot experience, Yvonne is an amazing resource. She can compare intelligently this method versus that and help you to understand all there is to know. And James is a super barefoot trimmer.

www.wholehorsetrim.com This is the website of Eddie Drabek, another one of my heroes. Eddie is a wonderful trimmer in Houston, Texas, and an articulate and inspirational educator and spokesman for getting metal shoes off horses. Read everything he has written, including the pieces on all the horses whose lives he has saved by taking them barefoot.

About the Author

Joe is the creator of the canine superstar Benji and the writer, producer, and director of all five Benji theatrical films and various television programs. He has authored eighteen books, including three novelizations of his own screenplays, ten horse-related books, and a book about his relationship with God. Joe spends time speaking around the country on behalf of kids, homeless pets, wild horses, and proper care for all horses. He has two sons, Joe Camp III and Brandon Camp, both movers and shakers in the movie business, and three stepchildren in college. He and his wife, Kathleen, eight horses, six chickens, five dogs, and a cat live in rural middle Tennessee. For more about Joe visit thesoulofahorse.com.

WHAT CRITICS ARE SAYING ABOUT
The Soul of a Horse

"Joe Camp is a master storyteller." —*New York Times*

"Joe Camp is a natural when it comes to understanding how animals tick and a genius at telling us their story. His books are must-reads for those who love animals of any species."
—Monty Roberts, author of the *New York Times* bestseller *The Man Who Listens to Horses*

"The tightly written, simply designed, and powerfully drawn chapters often read like short stories that flow from the heart. Camp has become something of a master at telling us what can be learned from animals, in this case specifically horses, without making us realize we have been educated, and, that is, perhaps, the mark of a real teacher." —Jack L. Kennedy, *Joplin Independent*

"One cannot help but be touched by Camp's love and sympathy for animals and by his eloquence on the subject." —Michael Korda, *Washington Post*

"Joe Camp is a gifted storyteller and the results are magical. Joe entertains, educates, and empowers, baring his own soul while articulating keystone principles of a modern revolution in horsemanship."
—Rick Lamb, author and TV/radio host of *The Horse Show*

FOR MORE INFORMATION
PLEASE VISIT ONE OF THESE WEBSITES:

thesoulofahorse.com

thesoulofahorse.com/blog

The Soul of a Horse Page on Facebook

The Soul of a Horse Channel on YouTube

Joe and The Soul of a Horse on Twitter

FOR KATHLEEN

WITHOUT HER LOVE THIS JOURNEY WOULD NOT HAVE EXISTED